Carbon Fibers and Their Composite Materials

Carbon Fibers and Their Composite Materials

Special Issue Editor

Luke Henderson

MDPI • Basel • Beijing • Wuhan • Barcelona • Belgrade

MDPI

Special Issue Editor
Luke Henderson
Deakin University, Australia

Editorial Office
MDPI
St. Alban-Anlage 66
4052 Basel, Switzerland

This is a reprint of articles from the Special Issue published online in the open access journal *Materials* (ISSN 1996-1944) from 2018 to 2019 (available at: https://www.mdpi.com/journal/materials/special_issues/c_fiber)

For citation purposes, cite each article independently as indicated on the article page online and as indicated below:

LastName, A.A.; LastName, B.B.; LastName, C.C. Article Title. *Journal Name* **Year**, *Article Number*, Page Range.

ISBN 978-3-03921-102-9 (Pbk)
ISBN 978-3-03921-103-6 (PDF)

Contents

About the Special Issue Editor

Luke Henderson obtained his Ph.D. from Griffith University in late 2007, working on the total synthesis of tetralone-derived natural products. A 12 month postdoc at Oxford University followed, working in the area of Pd-mediated cross-couplings for nitrogen heterocycle synthesis. He returned to Australia at Deakin University as an inaugural Alfred Deakin postdoctoral researcher in 2008. From this time, Luke has branched into different fields, leading to a major research focus on the manipulation of surface chemistry, primarily of carbon-based materials such as carbon fibers

Preface to "Carbon Fibers and Their Composite Materials"

The use of composites in modern infrastructure and transport has facilitated a huge decrease in fuel consumption and the corresponding emissions. The use of carbon fiber-reinforced polymer (CFRP) composites has been prolific in the aerospace and automotive industries, though typically reserved for low production volume and high value applications for the latter. While the promise of carbon fiber composites continues to be spoken of, the Achilles' heel of any composite material, including carbon fiber, is the interface and interphase region between dissimilar materials. In CFRPs, the ability of the polymer to transfer stress across the interface/interphase region to the strong reinforcing fibers is critical to the global performance of the part. This region has been under investigation and subjected to modification for decades with little meaningful progress in the large scale. In this collection of studies, a series of investigations is carried out in this area to contribute to the body of knowledge concerning this critical juncture.

Luke Henderson
Special Issue Editor

materials

MDPI

Article

Electrolytic Surface Treatment for Improved Adhesion between Carbon Fibre and Polycarbonate

Jan Henk Kamps [1,*], Luke C. Henderson [2], Christina Scheffler [3], Ruud van der Heijden [1], Frank Simon [3], Teena Bonizzi [4] and Nikhil Verghese [5]

1 SABIC, Plasticslaan 1, 4612PX Bergen op Zoom, The Netherlands; ruud.vanderheijden@sabic.com
2 Institute for Frontier Materials, Carbon Nexus, Deakin University, 75 Pigdons Rd,
   Waurn Ponds, VIC 3216, Australia; luke.henderson@deakin.edu.au
3 Leibniz-Institut für Polymerforschung Dresden e.V. (IPF), Hohe Straße 6, 01069 Dresden, Germany;
   scheffler@ipfdd.de (C.S.); frsimon@ipfdd.de (F.S.)
4 SABIC Technology Center, 6160 AL Geleen, The Netherlands; Teena.Bonizzi@sabic.com
5 SABIC Technology Center, Sugar Land, Houston, TX 77478, USA; nikhil.verghese@sabic.com
* Correspondence: janhenk.kamps@sabic.com; Tel.: +31-164-293-482

Received: 13 October 2018; Accepted: 6 November 2018; Published: 12 November 2018

Abstract: To achieve good mechanical properties of carbon fibre-reinforced polycarbonate composites, the fibre-matrix adhesion must be dialled to an optimum level. The electrolytic surface treatment of carbon fibres during their production is one of the possible means of adapting the surface characteristics of the fibres. The production of a range of tailored fibres with varying surface treatments (adjusting the current, potential, and conductivity) was followed by contact angle, inverse gas chromatography and X-ray photoelectron spectroscopy measurements, which revealed a significant increase in polarity and hydroxyl, carboxyl, and nitrile groups on the fibre surface. Accordingly, an increase in the fibre-matrix interaction indicated by a higher interfacial shear strength was observed with the single fibre pull-out force-displacement curves. The statistical analysis identified the correlation between the process settings, fibre surface characteristics, and the performance of the fibres during single fibre pull-out testing.

Keywords: carbon fibre; surface treatment; polycarbonate; composites; interfacial adhesion; single fibre pull out

---

1. Introduction

Improving mechanical properties through the addition of reinforcing fibres is a common approach used in a range of thermoplastic materials [1–4]. An important parameter faced in all research in this domain is the role of the interface between the fibre and the resin. To enable and exploit the mechanical property profile of fibre-reinforced thermoplastic composites, fibre-matrix adhesion must be at an optimum level [1–3]. An increase in the adhesion between carbon fibres and the polymer matrix can be achieved using different approaches, which are summarized in detailed review articles [5,6]. In general, wet-chemical (sizing/polymer finish, acidic modification, and electrochemical modification), dry-chemical (plasma/high energy irradiation modification, nickel surface coating, and thermal modification) and multiscale methods by applying nano-particles onto the surface are used to modify the carbon fibre surface. Each combination of fibre and matrix material will have its own ideal approach; for polycarbonate, specific studies have been conducted, mainly with respect to oxygen plasma-treated carbon fibres [7–9] or electrochemical oxidation [10–13], generally showing a significant increase in adhesion to polycarbonate after treatment. Table 1 gives an overview of the studies documented in literature and their results. The interfacial shear strength characterization can be approached using different micromechanical testing methods, each of which has its own

---

unique procedure [14–18]. In this study, we focused on the single fibre pull-out, which is very suitable for evaluating the interfacial shear strength on a microscopic scale involving viscous polymers like polycarbonate.

Combining the modification of carbon fibres with surface characterization, followed by single fibre pull-out testing gives a detailed insight into the crucial parameters that control the interface and its impact on composite performance. Linking this approach with statistical studies to demonstrate the value of surface treatment for interface formation and compatibility of specific process settings with polycarbonate completes the work documented here.

Table 1. Overview of the material modifications, processing conditions, and micromechanical tests applied on carbon fibre—polycarbonate composites to increase and characterize the fibre-matrix adhesion by the interfacial shear strength (IFSS); the results represent the lowest and highest achieved value of the investigated materials for each reference); * M_w = molecular weight, ** SD = standard deviation.

Fibre	Treatment	Matrix	Testing Method	IFSS ± SD **	Ref.
PAN-based unmodified, unsized CF (Idemitsu Kosan, Tokyo, Japan)	Anodic oxidation (electrolyte solution: K_2CO_3/KOH; KNO_3/KOH)	PC (Makrofol®, Bayer, Leverkusen, Germany)	Microdroplet pull-off test	9.6 ± 1.1 MPa (not oxidized); 14.7 ± 3.1 MPa (2.5 min in KNO_3/KOH)	[10]
PAN-based CF with unknown sizing (12K, HTS40, Toho Inc. Corp., Tokyo, Japan) and self-prepared CF with epoxy sizing	Electrochemical oxidation using a 0.1 mol/L NaOH electrolyte	PC (Dongguang Plastic Film Corporation, Dongguang, China), focusing on polycarbonate backbone transesterification	Single fibre fragmentation test	25.04 ± 1.08 MPa (not oxidized); 47.53 ± 1.23 MPa (15 min treatment time)	[11]
PAN-based unmodified (UT) and oxidized (ST) CF (Toray Industries Inc., Tokyo, Japan)	Electrochemical oxidation	Bisphenol-A based PC with varying M_w * PC1 M_w 25,000 g/mol PC2 M_w 32,000–36,000 g/mol (consolidation temperature 230–310 °C)	Single fibre fragmentation test	PC 1: (230/310 °C) UT: 30.2/41.0 MPa ST: 43.8/56.5 MPa PC 2: UT: 42.8/48.4 MPa ST: 59.3/67.9 MPa	[12]
UHM pitch-based CF; HT PAN-based CF; both untreated and unsized	Microwave O_2-plasma oxidation	PC Makrolon® 2805 (Bayer, Leverkusen, Germany)	Single fibre fragmentation test	HT: 24.0 ± 2 MPa HT-ox.: 27.7 ± 2 MPa UHM: 12.2 ± 1 MPa UHM-ox: 46.7 ± 3 MPa	[7]
PAN-based CF (Hexcel Magnamite® IM7, Stamford, CT, USA)	Commercial oxidative surface treatment at different grades	linear amorphous thermoplastic, Bisphenol-A based PC (GE Plastics, Inc., Pittsfield, MA, USA), M_w 31,000 g/mol	Microindentation test	100% ox.: 27.0 ± 1.9 MPa 400% ox.: 28.6 ± 3.2 MPa	[13]
PAN-based CF: Magnamite AS1 and AS4 (Hercules Aerospace, Wilmington, NC, USA)	Plasma treatment with ammonia, argon, nitrogen and oxygen	Polycarbonate LEXAN™ 101, (SABIC, Bergen op Zoom, The Netherlands)	Single fibre fragmentation test	lc/d 102% for ammonia treated lc/d 100% for ammonia treated lc/d 90% for argon treated lc/d 65% for oxygen treated	[8]
PAN-based CF, C320.00A, Sigri SGL Carbon, Wiesbaden, Germany	Low pressure oxygen plasma	PC Macrofol® DE 1-1 (Bayer AG, Leverkusen, Germany)	Single fibre fragmentation test	11.1 ± 1.2 MPa (no treatment) 9.8 ± 1.4 MPa (20 min treatment)	[9]
PAN-based CF, unsized	Commercial process, undisclosed	Functionalized polycarbonate (SABIC, Bergen op Zoom, The Netherlands)	Single fibre pull-out	33.9 ± 9.1 MPa (reference) 42.2 ± 9.0 MPa (functionalized PC)	[19]

2. Materials and Methods

The carbonization of poly (1-acrylonitrile) fibre, followed by varying surface treatments (adjusting the current, potential, and conductivity), created a range of seven different fibre samples used in this study (Table 2).

For the manufacture of carbon fibres in this study, three surface treatment variables were taken into account, which are thought to affect fibre-to-matrix adhesion. These include the current passed through the fibre during surface treatment, the potential applied to the fibre, and the conductivity of the electrolyte used in the bath (in this instance, ammonium hydrogen carbonate, $[NH_4]^+$ $[HCO_3]^-$).

It should be noted that due to the continuous nature of this production methodology, and the fact that the electrolyte is partially consumed during the surface treatment process, maintaining the exact level of each amperage, potential, and conductivity is challenging. Thus, different experiments used values as similar as possible for each of these, and these variables were classed into 'bands' of low, medium, and high for current (8, 14, and 26 A, respectively), and potential (5.7–5.8, 8–8.1, and 12.5–13.5 V, respectively). Only medium and high conductivities (17.0–17.5, and 31.2–31.4 mS/cm^{-1}, respectively) were investigated, as low conductivity of the electrochemical bath carries a risk of equipment malfunction or breakage.

To minimize the effect of unknown influences, a control sample (sample number 1), which did not have any surface treatment applied, was included in the experiment. This sample was collected directly after being passed through the high-temperature furnace.

Table 2. Surface treatment parameters and physical properties of resultant fibres.

Sample Number	Current (A)	Potential (V)	Conductivity (mS/cm)
1	-	-	-
2	8	5.8	17.5
3	14	8	17.5
4	26	13.5	17
5	26	12.5	31.3
6	14	8.1	31.4
7	8	5.7	31.2

LEXAN™ HF1110, a polycarbonate homopolymer (BPA) produced on a commercial scale by SABIC (Saudi Basic Industries Corporation, Riyadh, Saudi Arabia) and available as high-flow general-purpose grade, was used for the single fibre pull-out (SFPO) testing, selected for its lower viscosity, enabling efficient fibre embedding.

2.1. Inverse Gas Chromatography (IGC)—Surface Free Energy Analysis (SEA)

A series of *n*-alkanes (*n*-hexane, *n*-heptane, *n*-octane, and *n*-nonane) and polar probes (chloroform, ethyl acetate, 1,4-dioxane, ethanol, and dichloromethane) were injected into a column, which was filled with the fibre samples with specific fractional surface coverages, and their retention times were measured. The retention times (t) were converted into retention volumes. The dispersive free surface energy (γ_S^D) and specific free energy of desorption (ΔG_{SP}^0) values on the surface of the fibre samples were determined in accordance with the standard method described by Jones [20].

The (ΔG_{SP}^0) value obtained from the chloroform and ethyl acetate pair of mono-functional acidic and basic probes was used to determine the acid and base properties of the samples by applying an acid-base theory developed by van Oss [18,21]. The specific component of the surface free energy (γ_S^{AB}) was calculated for this acid (Lewis electron pair acceptor) base (Lewis electron pair donor) pair. The so-called term 'specific component of the surface free energy' is widely used. However, according to the definition of the surface free energy {($\partial f / \partial o$) = free energy (f), which is necessary to increase the surface (o)} it seems to be wrong to separate the surface free energy into dispersive and specific components because the surface free energy is an intrinsic value of the solid surface and does not depend on interacting liquids. The energy, which is determined from a solid surface coming into contact with a liquid, must be considered as interaction energy (both two phases contribute to the interaction energy). However, there is no problem in determining the interaction energy using IGC and splitting the interaction energy values into the contributions of dispersive (interaction) energy and specific (interaction) energy values.

The total surface free energy, γ_S^T, was calculated as the sum of the dispersive (γ_S^D) and specific (γ_S^{AB}) energy contributions. Fitting the data to an exponential decay function, $y = y_0 + A \exp[-x \cdot t^{-1}]$, allowed for extrapolation across the entire range (0–100%) of the surface coverage (x), where y_0 is the value of the function at infinity.

2.2. Tensile Testing

Bare fibre samples were tested using a Favimat+ Robot 2 single fibre tester (Textechno H. Stein, Mönchengladbach, Germany) which automatically records linear density and force extension data for individual fibres loaded into a magazine (25 samples) with a pretension weight of 80 ± 5 mg attached to the bottom of each carbon fibre. Linear density was recorded using a length of 25 mm and a tension of 0.15 mN (as per the supplier specifications). The tensile load extension curves were collected at 1.0 mm/min using a gauge length of 25 mm and a pretension of 1.0 cN/tex. The load data were normalized by dividing by the linear density to give the specific stress strain curves from which the tensile strength (ultimate specific stress or tenacity) and specific modulus could be determined.

2.3. Tensiometer: Contact Angle and Surface Free Energy

The contact angles (CA) of fibre samples 1–7 with deionized water (milli-Q) and 1-bromonaphthalene (97%, Sigma-Aldrich, Taufkirchen, Germany) were measured on a force tensiometer K100SF with LabDesk 3.2.2 software from KRÜSS GmbH (Hamburg, Germany), which was placed on a TS-150 LP dynamic antivibration system supplied by TABLE STABLE (Mettmenstetten, Switzerland). The measurements were performed at ambient conditions.

In each test, a force-displacement curve was recorded while immersing a fibre into one of the test liquids at a length of 5 mm with a speed of 3 mm/min and a data acquisition step of 0.02 mm. For the detection of the fibre (a sudden change in force), a detection speed of 6 mm/min was used and the detection sensitivity was set at 2×10^{-5}–7.5×10^{-5} g. By regression of the force-displacement curve and extrapolation to 0 mm immersion depth, the wetting force F was determined and the advancing contact angle (θ_a) was calculated with the Wilhelmy Equation:

$$\cos\theta_a = F/(L\sigma) \tag{1}$$

where σ is the total surface tension of the liquid (water = 72.8 mN/m, 1-bromonaphthalene = 44.6 mN/m) and L is the perimeter of the fibre based on the average fibre diameter determined for each sample during tensile testing (see Section 2.2).

For each sample, 10 fibres were tested per test liquid, resulting in an average θ_a per test liquid. The average θ_a with water and 1-bromonaphthalene were used to calculate the surface free energy (SFE) values of all the fibre samples, using the Owens, Wendt, Rabel, and Kaelble (OWRK) method [22]. The total SFE of each sample equals the sum of a polar (σ^P) and dispersive surface energy component (σ^D). The surface polarity was determined by taking the ratio—reflected as a percentage—of the polar component to the overall SFE.

2.4. Single Fibre Pull-Out Test (SFPO)

The interfacial adhesion strength between the fibre and matrix was evaluated by means of a SFPO using purpose-built embedding equipment constructed at IPF Dresden (Germany) [15,23]. Samples were prepared by accurately embedding one end of the selected single fibre in the matrix (perpendicularly) with a pre-selected embedding length l_e (l_e = 150 µm). For polycarbonate, an embedding temperature of 300 °C was required and embedding was carried out at controlled atmosphere and temperature. After embedding, the temperature was held at 300 °C for about 30 s, before cooling down to ambient temperature. The pull-out test was carried out with a force accuracy of 1 mN, a displacement accuracy of 0.07 µm, and a loading rate of 0.01 µm/s at ambient conditions (using a self-made pull-out apparatus). The force-displacement curves and the maximum force, F_{max}, required for pulling the fibre out of the matrix were measured. After testing, the fibre diameter, d_f, was measured using optical microscopy; l_e was determined using the force-displacement curve and cross-checked using scanning electron microscope (SEM) Ultra (Carl Zeiss AG, Oberkochen, Germany). The adhesion bond strength between the fibre and the matrix was characterized by the values of the apparent interfacial shear strength ($\tau_{app} = F_{max}/(\pi \times d_f \times l_e)$). Other interfacial parameters (such as local interfacial shear strength τ_d and interfacial frictional stress τ_f) were not considered in this work for analysing the fibre-matrix adhesion. Most of the curves did not follow the characteristic shape of the pull-out curve as described in Reference [18], meaning that the determination of the characteristic points for modelling (debonding force F_d, minimum force after debonding based on friction F_b) were not clearly identifiable [24,25]. Instead, the debonding work (from l_e = 0 to l_e at F_{max}) and pull-out work (from l_e at F_{max} to maximum l_e achieved at complete fibre pull-out) were used for comparison. Each fibre/matrix combination was evaluated in about 15–20 single tests. The filament surface before and after the pull-out test was evaluated using (SEM).

2.5. X-ray Photoelectron Spectroscopy (XPS)

All the XPS studies were carried out by means of an Axis Ultra photoelectron spectrometer (Kratos Analytical, Manchester, UK), equipped with a monochromatic Al Kα (1486.6 eV) X-ray source of 300 W at 15 kV. A hemispheric analyzer set to pass energy of 160 eV for wide-scan spectra and 20 eV for high-resolution spectra was used to determine the kinetic energy of the photoelectrons. The sample (carbon fibre tow) was mounted on a sample holder using adhesive tape so that the analyzed area was over an opening in the sample holder, enabling exposure to the X-ray source during measurement. Although the carbon fibres were electrically conductive, a low-energy electron source in combination with a magnetic immersion lens was employed to avoid electrostatic charging of the sample that can occur by fixing the fibres on the sample holder with the insolating adhesive tape. All the recorded peaks were shifted by the same value to set the C 1s component peak of the saturated hydrocarbons to 285.00 eV. The quantitative elemental compositions were determined from the peak areas using experimentally determined sensitivity factors and the spectrometer transmission function. Kratos spectra deconvolution software was applied to the high-resolution spectra and the spectrum background was subtracted according to Shirley. The free parameters of the component peaks were their binding energy (BE), height, full width at half maximum, and the Gaussian-Lorentzian ratio.

2.6. Statistical Evaluation

The testing results are reported as single values or mean ± standard deviation when multiple repeat evaluations of the fibre sample were conducted. Table 3 lists the average values, standard deviation, and sample size for (Favimat) tensile testing, and the single fibre pull-out measurements. In the results section the average values, standard deviation, and sample size for the contact angle measurements, which are used to calculate the energy and adhesion values listed in Table 3. Inverse gas chromatography was practiced on a bundle of fibres, resulting in responses based on the surfaces of numerous individual filaments; in all cases the line fit had a R^2 > 0.997, showing a good representation of the reported results. XPS was carried out by irradiating an area of approximately 3 mm^2 of the analyzed fibre bundles. From this irradiated area, the spectrometer collects nearly all the photoelectrons leaving the sample surface, measures their kinetic energy, and uses them to draw the corresponding spectrum, which reflects the average of the analyzed area, representing a large number of filaments.

To compare the physical properties of the six surface-treated samples to the control, the Dunnett's Method test was used, the Pearson correlation coefficient (r) was calculated, and the *p*-value for statistical significance was derived.

To evaluate the relationship between the IFSS and the surface treatment parameters (current, potential, and conductivity) a linear regression model was used with both univariate and multivariable results reported as a parameter estimate (95% confidence interval) with a *p*-value. The fit of the model was assessed visually, and no concerns were noted.

All the analyses were performed on JMP© Pro 13, SAS Institute Inc., Cary, NC, USA, and a *p*-value of less than 0.05 was considered as statistically significant.

Table 3. Complete overview of process settings and associated test results.

Sample Number	1	2	3	4	5	6	7	Testing Method
Current (A)	0	8	14	26	26	14	8	-
Potential (V)	0	5.8	8	13.5	12.5	8.1	5.7	-
Conductivity (mS/cm)	0	17.5	17.5	17	31.3	31.4	31.2	-
Elongation at Break (%)	1.58	1.60	1.63	1.64	1.70	1.79	1.63	Favimat
Standard deviation (n = 25)	0.24	0.29	0.24	0.28	0.32	0.23	0.24	-
Modulus (GPa)	259.85	261.44	266.24	262.06	261.81	263.09	264.22	Favimat
Standard deviation (n = 25)	3.59	4.58	11.36	3.26	5.20	4.73	4.26	-
Tensile strength (GPa)	3.84	3.88	4.05	4.02	4.13	4.38	4.00	Favimat
Standard deviation (n = 25)	0.61	0.72	0.62	0.71	0.80	0.58	0.62	-
Diameter (μm)	6.54	6.54	6.5	6.55	6.59	6.52	6.56	Favimat
Standard deviation (n = 25)	0.14	0.15	0.13	0.13	0.11	0.19	0.15	-
Total surface energy (mJ/m^2)	67.0	68.1	72.2	75.7	73.2	72.7	70.5	IGC
Dispersive surface energy (mJ/m^2)	51.9	47.4	46.1	47.8	46.9	46.4	47.4	IGC
Specific surface energy (mJ/m^2)	15.0	20.6	26.0	27.5	26.0	26.1	22.7	IGC
Atomic Conc. Hydroxyl (%)	1.50	1.90	2.15	3.68	3.24	3.36	3.10	XPS
Atomic Conc. Carboxyl (%)	1.10	1.51	1.62	2.93	3.05	2.15	1.80	XPS
Atomic Conc. Nitrile (%)	2.07	4.79	4.48	5.75	7.17	6.52	6.70	XPS
Total surface energy (mJ/m^2)	41.9	55.9	56.0	64.0	56.5	58.4	56.2	CA
Polar surface energy (mJ/m^2)	2.7	14.8	17.2	21.8	20.3	18.2	14.4	CA
Dispersive surface energy (mJ/m^2)	39.2	41.1	38.8	42.2	36.2	40.3	41.8	CA
Polarity (%)	6.5	26.4	30.7	34.0	35.9	31.1	25.6	CA
Adhesion energy ambient (mJ/m^2)	83.7	87.6	85.5	89.5	83.1	87.2	88.3	CA
Interfacial tension ambient (mN/m)	1.6	11.6	13.9	17.9	16.8	14.7	11.3	CA
Adhesion energy 260 °C (mJ/m^2)	50.6	71.0	72.6	78.8	74.6	74.4	70.9	CA
Interfacial tension 260 °C (mN/m)	19.4	13.0	11.4	13.3	9.9	12.2	13.4	CA
τ_{app} (N/mm^2)	48.8	50.1	55.2	43.2	54.7	49.5	33.3	SFPO
Standard deviation (n = 25)	12.4	14.0	11.5	11.1	6.5	18.9	15.1	-
W_{debond} (mN mm)	1.5	1.2	1.8	0.7	0.9	0.7	0.6	SFPO
Standard deviation (n = 25)	0.9	0.6	1.3	0.9	0.6	0.5	0.6	-
$W_{pullout}$ (mN mm)	2.1	1.6	1.5	2.9	2.2	1.3	2.0	SFPO
Standard deviation (n = 25)	0.9	0.6	0.7	2.7	2.8	0.7	1.1	-

3. Results

The work documented here spreads across different disciplines and techniques. An overview of the results of the production, fibre characterization, and single fibre pull-out testing are reported in Table 3.

3.1. Fibre Surface Treatment Results and Differences Observed

The characterization of the untreated fibres (sample 1) showed a tensile strength and Young's modulus of 3.84 and 259.85 GPa, respectively. For a comparison with a commercial product, these properties are slightly superior compared to automotive grade carbon fibres (T300, tensile strength and Young's modulus of 3.53 and 230 GPa, respectively). Samples 3 and 7 had a statistically significant increase in Young's modulus, though elongation at break and tensile strength were unchanged. Further improvements were observed when both the potential and current through the fibre were increased, at the same conductivity, though again, the only statistically significant change compared to sample 1 was with respect to the Young's modulus. Interestingly, further increasing the amperage and potential caused the Young's modulus to decrease slightly (Table 3, sample 4), and increasing the conductivity (Table 3, sample 5) corresponded to no meaningful property changes, suggesting that there is an optimal ratio and interplay between these three variables and that more of each, or even one, does not correspond to improved properties. Reverting to medium amperage and potential, which showed promise in sample 3, but increasing conductivity (Table 3, sample 6), had the most beneficial effects on the performance characteristics. All three measured parameters showed statistically significant changes relative to sample 1. Finally, combining low amperage and potential with high conductivity showed excellent improvements in all properties, suggesting that conductivity assists in the influence of the electrochemical treatments.

The application of current appeared to influence the modulus, with lower current settings being associated with a higher modulus. However, this effect can be modified by the potential setting. In particular, when the potential setting is low, increasing the current is associated with a higher modulus, but when the potential setting is high, increasing the current is associated with a lower modulus.

The fibre surface was examined to ensure no pitting or surface defects had arisen on the fibre surface due to these oxidative procedures. Given the tensile strength data acquired for these samples, it is unlikely that any defects had been introduced to the surface; nevertheless, imaging the fibre using SEM was undertaken in the interest of thoroughness (Figure 1).

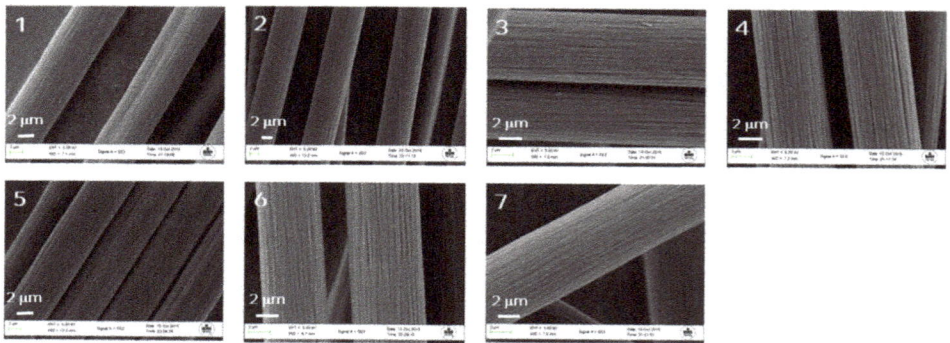

Figure 1. SEM images of all treated fibres from this study; sample 1 is the untreated sample, samples 2–7 show the same surface features and no surface defects have been detected.

The visual examination of the fibres displayed no obvious changes compared to sample 1, which had not undergone any surface treatment. The longitudinal striations and fibre diameter (approx. 7 μm) were observed with all samples, suggesting that the surface treatment conditions, while aggressive in some instances, did not result in substantial fibre degradation. Given the consistency in surface structure and morphology, we turned our attention next to the examination of the surface chemistry using Inverse Gas Chromatography (IGC).

3.2. Treatment Impact on Surface Energy and Functional Groups, Matching with PC

To determine the nature of the acid/base and the dispersive surface energies of the treated carbon fibres, we used IGC. In this technique, a column filled with the carbon fibres is injected with gaseous probes, which interact with various functional groups on the surface of the fibres. Typically, a range of non-polar (*n*-alkanes) and polar (ethyl acetate, ethanol, etc.) test liquids are used to determine the dispersive and Lewis acid/base properties, respectively (Table 4 and Figure 2).

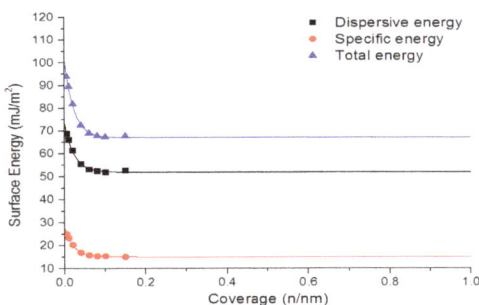

Figure 2. Inverse gas chromatography results (sample 1).

Given the nature of this technique, a comparison of the absolute values is not informative, therefore the ratio of dispersive and polar energies is provided to give a more meaningful comparison between the samples. Sample 1, as expected, possessed a very high dispersive energy component, resulting from the highly graphitic nature of this fibre.

There is some evidence to suggest that increasing the current will decrease the dispersive surface energy and increase the specific surface energy, as can be observed in Table 4. Furthermore, the potential applied is likely to have a modifying effect on the surface properties. The similarity of the specific energy values for samples 3 and 4 is counter-intuitive considering that the oxidative treatment was more aggressive for sample 4 than for sample 3, suggesting that a plateau was reached under these conditions, perhaps dictated by the concentration, and thus conductivity, of the electrolyte.

Table 4. IGC results of the produced fibres.

Sample Number	Dispersive Energy (mJ/m^2)	Specific (Acid-Base) (mJ/m^2)	Total (mJ/m^2)	Ratio of Dispersive and Specific Energies [a]
1	51.94 (77.5%)	15.04 (22.5%)	66.98	3.45:1.0
2	47.41 (69.8%)	20.55 (30.2%)	68.14	2.31:1.0
3	46.06 (63.9%)	26.02 (36.1%)	72.19	1.77:1.0
4	47.83 (63.5%)	27.45 (36.5%)	75.72	1.74:1.0
5	46.88 (64.4%)	25.97 (35.6%)	73.23	1.81:1.0
6	46.40 (64.0%)	26.09 (36.0%)	72.65	1.78:1.0
7	47.38 (67.5%)	22.73 (32.5%)	70.49	2.08:1.0

[a] Determined by dispersive/specific energies.

A similar observation can be made when examining samples 5 and 6, where the polar portion of the surface energy remains at approximately 35–36% of the total surface energy, again suggesting a plateau of oxidative treatment and installation of polar functional groups. Interestingly, sample 7 shows a distinct decrease in polar surface energy (32%), relative to the other oxidized samples, which corresponds to a decrease in both current and applied potential.

While IGC thermodynamically described the interactions of solid surfaces to the probe molecules in their environment, XPS offered the opportunity to analyze the type and number of functional groups in the surface region of the differently treated carbon fibres. The wide-scan XPS spectra (Figure 3, left column) showed—with the exception of hydrogen—all the elements in the surface region of the carbon fibres. Besides the metal ions, such as sodium, magnesium, silicon, and calcium that occur only as traces (regarding carbon content, their contents were less than 0.5 at-%), considerable amounts of nitrogen and oxygen were detected on the carbon fibre surface.

Although oxygen may also be bonded in counter ions of the metal ions, it can be assumed that the majority of the oxygen atoms were covalently bonded to the carbon fibres. Nitrogen, which was found on the carbon fibre surfaces, could be a constituent of functional surface groups but also a residue of the ammonium salt (NH$_4^+$), which was used during the electrical oxidation process. Shape-analysis of the high-resolution element spectra is an established method to study the different binding states of the atoms in the surface region of solids. However, due to the so-called 'shake-up' phenomena, which were observed in XPS spectra recorded from substances consisting of graphite-like lattices, such as carbon fibres, the deconvolution of the C 1s spectra is generally difficult. Graphite-like lattices consist of sp^2-hybridized carbon atoms in which the π-bonded p$_z$-electrons can be extensively delocalized. Each linear combination of two of the p$_z$ wave functions gives wave functions of one π-orbital occupied by the two p$_z$-electrons and one unoccupied π*-orbital. In the case of graphite-like structures, the high number of possible linear combinations leads to a quasicontinuum of energy levels that can be occupied by electrons. Energy from an external source can be consumed to transfer a p$_z$-electron from its π-orbital (ground state) into a π*-orbital (excited states). The C 1s spectra shows the photoelectrons emitted from the ground as well as excited states. The latter contribute to the shake-up peaks mentioned above.

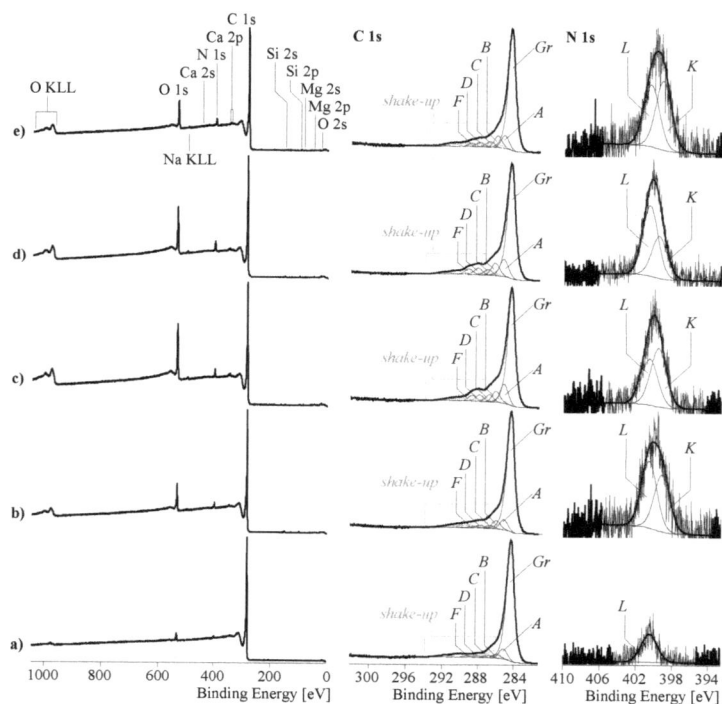

Figure 3. Wide-scan photoelectron spectra (left column), C 1s (middle column) and N 1s high-resolution photoelectron spectra (right column) recorded from unmodified carbon fibres (sample 1) (**a**), and electro-chemically modified carbon fibres at low current and low current and low conductivity (sample 2) (**b**), high current and low conductivity (sample 4) (**c**), high current and high conductivity (sample 5) (**d**), and low current and high conductivity (sample 7) (**e**).

The C 1s spectra recorded from all the carbon fibre samples are characterized by intense shake-up peaks appearing at binding energy values higher than 286 eV (Figure 3, middle column). In the same region, component peaks identifying different functional groups were expected. In order to separate the shake-up peaks overlapping the component peaks, it was assumed that the different surface modifications had the same effect on the π → π* transition probabilities and thus on the shape and intensities of all the shake-up peaks. The C 1s peak areas remained after subtraction, and the shake-up peaks were deconvoluted into six component peaks, showing different binding states of carbon. The most intense component peaks Gr (at 284.14 eV) resulted from the photoelectrons escaped from the sp^2-hybridized carbon atoms, forming the graphite-like lattice of the carbon fibre material. Saturated hydrocarbons in the sp^3 hybrid state, which did not have heteroatoms as binding partners, were assigned as component peaks A (at 285.00 eV). The presence of saturated hydrocarbons is frequently observed in surface analysis because non-specifically adsorbed contaminations mainly consist of alkanes and their derivatives. Component peaks B (at 285.84 eV) show C–N bonds of amines, C=N bonds of imines, and/or C≡N of nitrile groups. Surprisingly, the intensities of all component peaks B ($[B]$) were significantly higher than the [N]:[C] ratios independently determined from the wide-scan spectra ($[B] \approx 1.6$ [N]:[C]). Obviously, considerable amounts of the nitrogen atoms were present as bound to two carbon atoms, which is well-known from the oxidized cyclization of the PAN structure before the carbonization process of the fibres [26]. The introduction of oxygen in the surface region of the carbon fibre samples resulted in the formation of C–O bonds of mainly phenolic groups (component peaks C at 286.69 eV), quinone-like groups (C=O as component peaks D at 287.77 eV), and carboxyl groups (O=C–OH) and their corresponding carboxylates ($^-$O–C=O ↔ O=C–O$^-$) both as component peaks F at 288.72 eV. Table 5 summarizes the fractions of the component peak areas and thus gives an overview of the number of different functional groups.

Table 5. Fractions of component peak areas.

Sample Number	[N]:[C]	[O]:[C]	[B]	[C]	[D]	[F]
1	0.011	0.022	0.021	0.015	0.008	0.011
2	0.030	0.084	0.048	0.019	0.017	0.015
3	0.028	0.105	0.045	0.022	0.021	0.016
4	0.036	0.163	0.058	0.037	0.034	0.029
5	0.045	0.142	0.072	0.032	0.038	0.031
6	0.042	0.107	0.065	0.034	0.032	0.022
7	0.042	0.087	0.067	0.031	0.025	0.018

While the N 1s spectrum recorded from the unmodified carbon fibres showed a unimodal distribution of the photoelectrons around the peak maximum at 400.67 eV, the N 1s spectra of electro-chemically treated were deconvoluted into two component peaks, K and L. According to the binding energy values found for component peaks L (400.22 eV), it was assumed that these component peaks appeared from protonated nitrogen species, such as adsorbed ammonium ions (NH_4^+) or protonated amino groups (C–N$^+$H). The component peaks K were found at about 399 eV, which is a very small value for organically bonded nitrogen. The chemical shift to small binding energy values indicated increased electron densities at the nitrogen atoms probably caused by C=N bonds of azoles [27] or azabenzenes in the immediate environment of highly conjugated π-electron systems, for example. In contrast, the binding energy for the triply bonded nitrogen in the nitrile groups (C≡N) is expected at 399.5 eV [28].

As H-acidic compounds, phenol and carboxyl groups are Brønsted and Lewis acids. Their deprotonated species, the phenolate and carboxylate ions, act as Brønsted and Lewis bases. Brønsted basic amino groups can be protonated by hydronium ions (H_3O^+). Amino, azole, and azabenzene groups belong to the group of nitrogen bases. Due to the $-I$ effect of the nitrogen atom and the ability of nitrogen to bind a proton via its free electron pair, the nitrile group has an ambidentate character.

Contact angle measurements with a single fibre tensiometer resulted in a total surface free energy (SFE) and a surface polarity, which is the percentage of the total SFE that is due to the polar surface energy component, of the tested fibres (Table 6). All the surface-treated fibres had a numerically higher total SFE compared to the untreated fibre (sample 1, 41.9 mJ/m^2), with fibre sample 4 having the highest value, 64.0 mJ/m^2. In addition, all the surface-treated fibres had a higher surface polarity than the untreated fibre, and increasing the potential was associated with increasing polar surface energy.

From the SFE values, a wetting envelope (Figure 4) for complete wetting could be calculated, which describes all the combinations of the polar (y-axis) and dispersive (x-axis) surface tensions of a liquid that would result in a θ_a of $0°$ by solving the OWRK equation. These wetting profiles allow for the prediction of the wetting behaviour of the fibres: the combinations inside the envelope will result in complete wetting ($\theta_a = 0°$), while the combinations outside the envelope will not ($\theta_a > 0°$). Figure 4 shows the wetting envelopes for the untreated fibre (sample 1) and the extremes of the treated fibres (samples 4 and 7). It can be seen that, theoretically, improved wetting can be expected of the surface-treated fibres with commercial LEXAN™ HF1110 polycarbonate at both ambient temperature ($\sigma^P = 0.2$ mJ/m^2, $\sigma^D = 43.2$ mJ/m^2) and 260 °C ($\sigma^P = 19.9$ mN/m, $\sigma^D = 8.2$ mN/m) compared to the untreated fibres.

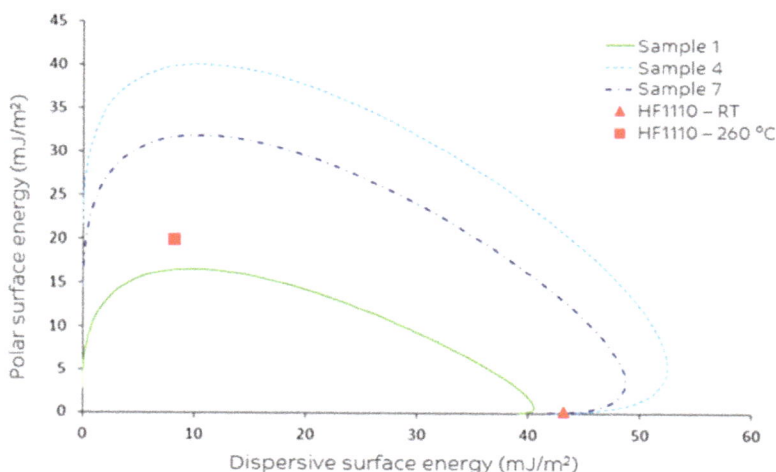

Figure 4. Wetting envelope for complete wetting ($\theta_a = 0°$).

Table 6. Tensiometer results of the fibres produced.

Sample Number	θ_a [Water] [a] (°)	θ_a 1-[Bromonaphthalene] [a] (°)	Total SFE (mJ/m^2)	Surface Polarity (%)
1	82.6 ± 3.2 [b]	28.7 ± 5.1	41.9	6.5
2	54.7 ± 3.9	22.2 ± 6.9	55.9	26.4
3	52.2 ± 3.8	29.1 ± 8.2	56.0	30.7
4	41.4 ± 3.7	17.8 ± 7.1	64.0	34.0
5	48.9 ± 4.5	35.8 ± 8.8	56.5	35.9
6	49.3 ± 4.4	24.4 ± 8.7	58.4	31.1
7	54.9 ± 4.5	19.7 ± 6.4	56.2	25.6

[a] Based on 10 measurements. [b] Based on 8 measurements.

To further quantify the compatibility with polycarbonate, the adhesion energy (ψ) and interfacial tension (γ) were calculated with the Fowkes/Dupre and Good's expression, respectively, using the SFE values of the fibres and commercial LEXAN™ HF1110 [29]. The adhesion energy describes how energetically favourable the initial formation of an interface is, whereas the interfacial tension describes the tendency of the formed interface to break in the future upon stress. For good interfacial properties, high adhesion energy and low interfacial tension are targeted. Although it is assumed that the SFE values of the fibres are not dependent on the temperature, the total SFE and surface polarity of the polycarbonate matrix material changes when transitioning from a solid at ambient temperature to a molten polymer at 260 °C. Therefore, the interfacial parameters and trends amongst the fibre samples depend on the conditions used to combine the materials. Assuming a melt impregnation process, all surface-treated fibres show improved adhesion energy compared to the untreated fibre (50.6 mJ/m^2), with the highest value being for fibre sample 4 (78.8 mJ/m^2), which had the highest total SFE and surface polarity (Table 7). The conclusion as to how surface treatment influenced interfacial tension depends highly on the temperature studied: at ambient temperature, the untreated fibre looks superior, whereas at 260 °C all treated fibres are better than the untreated fibre. Single fibre pull-out testing has been attempted next to give more clarity as to which parameters and conditions are most indicative for optimal interfacial adhesion.

Table 7. Adhesion energy and interfacial tension for LEXAN™ HF1110 polycarbonate.

Sample Number	Adhesion Energy (mJ/m^2)		Interfacial Tension (mN/m)	
	Ambient	260 °C	Ambient	260 °C
1	83.7	50.6	1.6	19.4
2	87.6	71.0	11.6	13.0
3	85.5	72.6	13.8	11.4
4	89.5	78.8	17.9	13.3
5	83.0	74.6	16.8	9.9
6	87.2	74.4	14.7	12.2
7	88.3	70.9	11.3	13.4

3.3. Single Fibre Pull-Out Test (SFPO)

Contact angle, IGC, and XPS measurements revealed a significant increase in the polarity and functional groups on the fibre surface due to the surface treatment. Accordingly, an increase in the fibre-matrix interaction indicated by higher maximum forces was observed using the SFPO force-displacement curves (Figure 5, showing three selected samples: untreated sample 1, highest (sample 3), and lowest (sample 7) maximum pull-out forces).

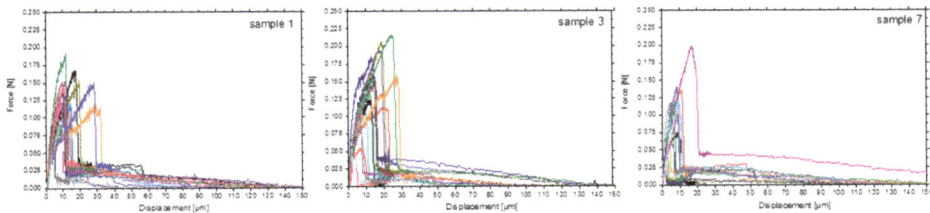

Figure 5. Force-displacement curves of sample 1 (no surface treatment), sample 3 (medium current, medium conductivity) as the most extensive, and sample 7 (low current, high conductivity) with the lowest fibre-matrix interaction, respectively.

However, it should be noted that even the untreated fibre already reveals a good interaction between the fibre and the PC matrix. To some extent, this might be due to the fact that the PC matrix near the fibre is considerably deformed (stretched) during the pull-out. Figure 6 presents the deformation of the meniscus on the fibre surface as well as the strong deformation of the matrix material near to the fibre entry point. Besides the contribution of the crack that is growing along the fibre surface during pull-out and the friction between the already debonded surface areas, matrix deformation also contributes to the maximum force achieved. This would explain the high forces in the case of untreated sample 1.

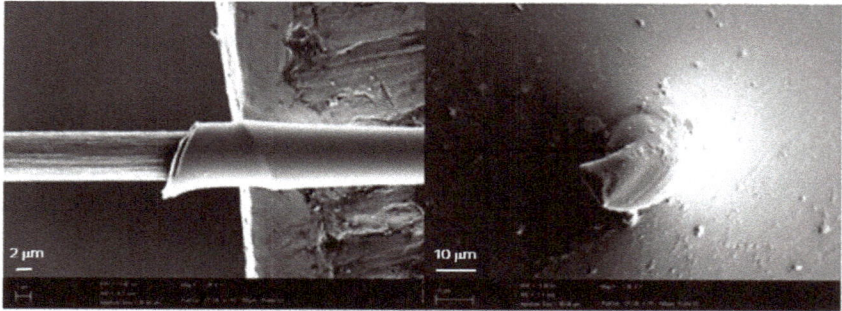

Figure 6. Stretching/yielding of the meniscus after pull-out testing; SEM image of the remaining part on the pulled-out fibre (**left**) and the stretched area of the fibre entry point in the PC droplet (**right**).

On the contrary, this kind of meniscus stretching also occurred for sample 7, which revealed not only the lowest values of τ_{app} and W_{debond} but also a drop in polar surface; these might be related to each other. The currently known models (stress-controlled model; energy-based model, model of adhesion pressure [25]) used to calculate the interfacial parameters (ultimate interfacial shear strength τ_{ult}, critical energy release rate G_{ic}) do not involve this kind of meniscus deformation. As mentioned in Section 3.4, the apparent shear strength τ_{app} as well as debonding and pull-out work were used to describe the fibre-matrix interaction for that reason. In general, increased shear strength τ_{app} was found for the treated samples; however, the measurements were accompanied by high scatter due to the non-circular fibre shape.

3.4. Correlations

A statistical analysis was carried out to find the correlation between the process parameters, surface characterization techniques, and SFPO results and their significance (Tables 8 and 9).

Table 8. Correlation factors on pair-wise comparisons between results, where 1 is the total positive linear correlation, 0 is no linear correlation, and −1 is the total negative linear correlation. The underlined factors show *p*-values < 0.05 and are considered as significant.

Factors	Elongation at Break	Modulus	Tensile Strength	Total Surface Energy	Dispersive Surface Energy	Atomic Conc. Hydroxyl	Atomic Conc. Carboxyl	Atomic Conc. Nitrile	Total Surface Energy	Polar Surface Energy	Dispersive Surface Energy	Polarity	Interfacial Tens. (Ambient)	Interfacial Tens. (260 °C)	τ_{app}	W_{debond}	$W_{pullout}$
	Favimat			IGC		XPS			CA						SFPO		
Current (A)	0.46	0.17	0.61	0.92	−0.56	0.77	0.96	0.67	0.77	0.89	−0.21	0.86	−0.73	0.20	−0.40	0.45	0.46
Potential (V)	0.47	0.26	0.57	0.93	−0.66	0.80	0.93	0.73	0.87	0.95	−0.09	0.92	−0.79	0.13	−0.44	0.39	0.47
Conductivity (mS/cm)	0.72	0.45	0.16	0.51	−0.79	0.73	0.56	0.96	0.64	0.70	−0.06	0.75	−0.80	−0.18	−0.66	−0.21	0.72
Elongation at Break	-	0.20	−0.09	0.54	−0.56	0.66	0.52	0.67	0.45	0.55	0.06	0.56	−0.55	0.18	−0.50	−0.32	1.00
Modulus	0.20	-	−0.39	0.38	−0.73	0.21	0.01	0.35	0.43	0.44	−0.17	0.48	−0.56	−0.04	0.13	−0.39	0.20
Tensile strength	0.36	0.38	-	0.60	−0.63	0.67	0.50	0.66	0.50	0.59	0.01	0.60	−0.59	0.17	−0.44	−0.34	0.98
Total surface energy (IGC)	0.54	−0.73	0.60	-	−0.60	0.86	0.87	0.66	0.83	0.88	0.02	0.83	−0.65	0.01	−0.44	0.39	0.54
Dispersive surface energy (IGC)	−0.56	0.21	−0.63	−0.60	-	−0.53	−0.45	−0.76	−0.79	−0.85	0.17	−0.90	0.94	−0.14	0.24	0.35	−0.56
Atomic conc. Hydroxyl (XPS)	0.66	0.01	0.67	0.86	−0.53	-	0.86	0.85	0.80	0.79	−0.17	0.74	−0.57	−0.33	−0.83	0.37	0.66
Atomic conc. Carboxyl (XPS)	0.52	0.35	0.50	0.87	−0.45	0.86	-	0.74	0.72	0.82	−0.01	0.79	−0.64	0.04	−0.60	0.53	0.52
Atomic conc. Nitrile (XPS)	0.67	0.43	0.66	0.66	−0.76	0.85	0.74	-	0.76	0.81	0.34	0.83	−0.81	−0.22	−0.76	0.04	0.67
Total surface energy (CA)	0.45	0.44	0.50	0.83	−0.79	0.80	0.72	0.76	-	0.95	0.03	0.91	−0.76	−0.15	−0.56	0.19	0.45
Polar surface energy (CA)	0.55	0.06	0.59	0.88	−0.85	0.79	0.82	0.81	0.95	-	-	0.99	−0.89	0.08	−0.46	0.15	0.55
Dispersive surface energy (CA)	−0.20	0.48	−0.17	0.01	0.02	0.17	−0.17	−0.01	0.34	0.03	-	−0.06	0.25	−0.73	−0.39	0.16	−0.20
Polarity (CA)	0.56	0.42	0.60	0.83	−0.90	0.74	0.79	0.83	0.91	0.99	−0.06	-	−0.95	0.14	−0.41	0.05	0.56
Interfacial tension (ambient) (CA)	0.55	−0.56	0.59	0.89	−0.84	0.79	0.83	0.81	0.94	1.00	0.01	0.99	-	0.10	−0.46	0.16	0.55
Interfacial tension (260 °C) (CA)	−0.55	−0.56	−0.59	−0.65	0.94	−0.57	−0.64	−0.81	−0.76	−0.89	0.25	−0.95	−0.89	-	0.26	0.18	−0.55

Table 9. *p*-values for the correlation factors on pair-wise comparisons; *p*-values < 0.05 are considered as significant and are underlined.

Factors	Elongation at Break	Modulus	Tensile Strength	Total Surface Energy	Dispersive Surface Energy	Atomic Conc. Hydroxyl	Atomic Conc. Carboxyl	Atomic Conc. Nitrile	Total Surface Energy	Polar Surface Energy	Dispersive Surface Energy	Polarity	Interfacial Tens. (Ambient)	Interfacial Tens. (260°)	τ_{app}	W_{debond}	$W_{pullout}$
	Favimat			IGC		XPS			CA						SFPO		
Current (A)	0.30	0.72	0.29	0.00	0.19	0.04	0.00	0.10	0.04	0.01	0.65	0.01	0.01	0.06	0.67	0.38	0.31
Potential (V)	0.29	0.58	0.27	0.00	0.10	0.03	0.00	0.06	0.01	0.00	0.85	0.00	0.00	0.04	0.79	0.32	0.39
Conductivity (mS/cm)	0.07	0.31	0.07	0.24	0.03	0.06	0.19	0.00	0.12	0.08	0.90	0.05	0.08	0.03	0.70	0.11	0.65
Elongation at Break	-	0.67	0.00	0.21	0.19	0.11	0.23	0.10	0.31	0.20	0.66	0.19	0.20	0.20	0.93	0.25	0.49
Modulus	0.67	-	0.42	0.41	0.06	0.65	0.99	0.44	0.33	0.32	0.90	0.27	0.34	0.19	0.72	0.79	0.39
Tensile strength	0.00	0.42	-	0.15	0.13	0.10	0.26	0.11	0.26	0.17	0.71	0.15	0.17	0.17	0.98	0.33	0.45
Total surface energy (IGC)	0.21	0.41	0.15	-	0.16	0.01	0.01	0.05	0.02	0.01	0.98	0.01	0.02	0.11	0.77	0.32	0.39
Dispersive surface energy (IGC)	0.19	0.06	0.13	0.16	-	0.22	0.31	0.01	0.03	0.02	0.96	0.02	0.00	0.18	0.47	0.60	0.44
Atomic conc. Hydroxyl (XPS)	0.11	0.65	0.10	0.01	0.22	-	0.01	0.06	0.07	0.02	0.71	0.01	0.02	0.12	0.93	0.02	0.41
Atomic conc. Carboxyl (XPS)	0.23	0.99	0.26	0.01	0.31	0.01	-	0.05	0.05	0.02	0.71	0.04	0.02	0.03	0.63	0.15	0.22
Atomic conc. Nitrile (XPS)	0.10	0.44	0.11	0.11	0.05	0.01	0.06	-	0.00	0.03	0.46	0.02	0.03	0.05	0.74	0.05	0.93
Total surface energy (CA)	0.31	0.33	0.26	0.02	0.03	0.03	0.07	0.05	-	0.03	0.95	0.03	0.00	0.01	0.87	0.20	0.69
Polar surface energy (CA)	0.20	0.32	0.17	0.01	0.02	0.03	0.02	0.03	0.00	-	0.89	0.00	0.00	0.59	0.06	0.30	0.76
Dispersive surface energy (CA)	0.66	0.90	0.71	0.98	0.96	0.71	0.71	0.99	0.46	0.95	-	0.89	0.99	0.00	0.77	0.39	0.73
Polarity (CA)	0.19	0.27	0.15	0.02	0.01	0.06	0.04	0.02	0.00	0.00	0.89	-	0.00	0.01	0.84	0.36	0.91
Interfacial tension (ambient) (CA)	0.20	0.34	0.17	0.01	0.02	0.03	0.02	0.03	0.00	0.00	0.99	0.00	-	0.01	0.77	0.30	0.73
Interfacial tension (260 °C) (CA)	0.20	0.19	0.17	0.11	0.00	0.18	0.12	0.03	0.05	0.01	0.59	0.00	0.01	-	0.57	0.57	0.70

4. Discussion

Several approaches to modifying carbon fibre surfaces can be followed and their impact on adhesion to polycarbonate is studied, as listed in Table 1. In addition to the differences in testing methods, as described in the introduction, different suppliers and grades of polycarbonates were also used, where the difference in molecular weight will impact viscosity (and therefore wetting/impregnation) and the data comparison. This study focused on the electrolytic surface treatment of carbon fibre during its production process. Improvements in interfacial shear strength of comparable approaches have been documented [7–13], and fragmentation tests or indentation were used to quantify the impact. The authors of this article have also investigated the modification of polycarbonate as a means to achieving improved adhesion [19], where an apparent interfacial shear strength of 33.9 MPa was found for the reference polycarbonate (HF1110). By using functional groups, the adhesion was improved to 42.2 MPa, in combination with a commercially available carbon fibre (unsized), without further information on the specific process parameters of production.

By controlling the process parameters of the electrolytic surface treatment, a range of samples were created, which were characterized using chemical and mechanical characterization techniques, to evaluate the impact of treatment as well as the predictability of interfacial shear strength.

The mechanical properties of the fibres were in all cases equal or superior to untreated sample 1, showing a slight increase in Young's modulus for samples 4 and 7, but no detrimental impact of the treatments were found and no obvious changes compared to sample 1 were observed from the SEM analysis.

From the inverse gas chromatography data, it would seem that the introduction of polar groups onto the surface of carbon fibres correlates well with the application of potential and current, which is in line with the observations made by contact angle measurements as well as the significant increase in the functional groups on the fibre surface observed using XPS.

There are challenges in controlling the exact level of amperage, potential, and conductivity applied to the samples, and when this is combined with the complexity of the analytical tests performed and a small sample size, establishing clear relationships between the process settings and the fibre characteristics was always going to be a challenge. However, we were able to demonstrate that current and potential are associated with a number of fibre features. In particular, we found that to influence τ_{app}, there is an important interplay between the current and potential settings which means that tailoring these settings is not straightforward but that it could be possible to use this knowledge to target particular applications.

While significant correlations were found between the fibre characteristics, we did not find a direct correlation between process settings, tensile strength measurement, inverse gas chromatography or contact angle results and the single fibre pull-out parameters. This could be due to the limitations of the correlation test, assuming the relationships are linear. However, at this stage we were unable to find a test that correlates well with τ_{app}, leaving the single fibre pull-out test as the most important analytical technique used in this study to predict interfacial shear strength.

The XPS characterization results did correlate significantly with the SFPO results. All the other techniques showed correlations among each other, but this did not render SFPO results (the most time-consuming and specialized technique) predictable enough for it to be acceptable to depend on it for research screening.

Further statistical evaluation of the presented dataset resulted in the predictive model for IFSS based on surface treatment process variables:

$$\tau_{app} = (-0.32 \times V \times I) + (0.24 \times V \times C) + (2.1856 \times V) + (2.4512 \times I) - (2.4084 \times C) + 48.7663 \quad (2)$$

where V is the voltage applied, I is the value of current applied to the electrolytic solution, and C is the value of conductivity of the electrolytic solution. Using this formula to calculate the IFSS makes it possible to select the right process settings targeting a specific value.

Verification of this model, using the predicted values based on the process settings used in the preparation of samples 1 to 7 versus the actual measured values, shows a very good correlation ($R^2 = 0.99$, $p = 0.0255$), as presented in Figure 7.

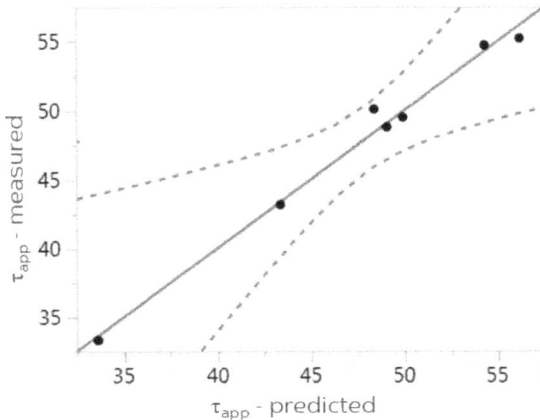

Figure 7. Linear correlation between τ_{app} predicted (Equation 2) and τ_{app} measured (samples 1–7).

5. Conclusions

The impact of the electrolytic oxidation of carbon fibre on adhesion to polycarbonate has been studied and the impact of the variation of process parameters discussed. A set of on-purpose fibre samples were produced and characterized with a range of surface characterization techniques (IGC, XPS, CA), and single fibre pull-out testing was used for the quantification of the interfacial shear strength between the fibre and the polycarbonate matrix.

The statistical analysis showed significant correlations between IGC, XPS, and CA, but no predictive model was found in the pair-wise comparison between the surface characterization results and the single fibre pull-out measurements.

The dataset produced resulted in a predictive model for interfacial shear strength based on the process parameters used for electrolytic oxidation of the carbon fibre. This model makes it possible to target a certain interfacial shear strength, as desired or specified for the carbon fibre-polycarbonate composite.

6. Patents

The results of this study are documented in patent application "Methods for electrolytic surface treatment of carbon fibers", USPTO serial number 62539879, published on IP.com with reference number IPCOM000252191D.

Author Contributions: This article is the result of collaboration between the authors, where J.H.K. and N.V. led the efforts and arranged all the contacts and connections. L.H. provided the settings for the fibre surface treatment and produced the samples used, including the IGC and mechanical performance. R.v.d.H. and J.H.K. oversaw the practicalities of the project, characterized the surface by tensiometer and collected all the input for further analysis. C.S., responsible for the coordination at IPF Dresden, provided the SFPO results and F.S. characterized the samples by XPS. T.B. used her analysis tools for the statistical analysis and for defining the correlations between the SFPO and the process parameters. All the authors contributed to the drafting of this document.

Acknowledgments: The work described in this document is the result of a joint effort from the authors, the SABIC Technology department, Carbon Nexus (Geelong, Australia) and contributors at the Leibniz-Institut für Polymerforschung Dresden e.V. (IPF). Special acknowledgement goes to the contributions of Reema Sinha, Haimanti Datta, Anton Kumanan, Mathilde Delory and Alexander van Goudswaard. Alma Rothe and Steffi Preßler are acknowledged for technical support with micromechanical testing.

Conflicts of Interest: The authors declare no conflict of interest.

References

1. Cogswell, F.N. *Thermoplastic Aromatic Polymer Composites: A Study of the Structure, Processing and Properties of Carbon Fibre Reinforced Polyetheretherketone and Related Materials*; Elsevier: Amsterdam, The Netherlands, 2013; ISBN 9781483164762.

2. Fu, S.Y.; Lauke, B.; Mäder, E.; Yue, C.Y.; Hu, X. Tensile properties of short-glass-fibre-and short-carbon-fibre-reinforced polypropylene composites. *Compos. Part A Appl. Sci. Manuf.* **2000**, *31*, 1117–1125. [CrossRef]

3. Botelho, E.C.; Rezende, M.C.; Lauke, B. Mechanical behavior of carbon fibre reinforced polyamide composites. *Compos. Sci. Technol.* **2003**, *63*, 1843–1855. [CrossRef]

4. Karger-Kocsis, J.; Mahmood, H.; Pegoretti, A. Recent advances in fibre/matrix interphase engineering for polymer composites. *Prog. Mater. Sci.* **2015**, *73*, 1–43. [CrossRef]

5. Sharma, M.; Gao, S.; Mäder, E.; Sharma, H.; Wei, L.Y.; Bijwe, J. Carbon fibre surfaces and composite interphases. *Compos. Sci. Technol.* **2014**, *102*, 35–50. [CrossRef]

6. Yao, S.S.; Jin, F.-L.; Rhee, K.Y.; Hui, D.; Park, S.J. Recent advances in carbon-fibre-reinforced thermoplastic composites: A review. *Compos. Part B Eng.* **2018**, *142*, 241–250. [CrossRef]

7. Montes-Morán, M.A.; Martínez-Alonso, A.; Tascón, J.M.D.; Paiva, M.C.; Bernardo, C.A. Effects of plasma oxidation on the surface and interfacial properties of carbon fibres/polycarbonate composites. *Carbon* **2001**, *39*, 1057–1068. [CrossRef]

8. Bascom, W.D.; Chen, W.-J. Effect of Plasma Treatment on the Adhesion of Carbon Fibres to Thermoplastic Polymers. *J. Adhes.* **1991**, *34*, 99–119. [CrossRef]

9. Bismarck, A.; Richter, D.; Wuertz, C.; Kumru, M.E.; Song, B.; Springer, J. Adhesion: Comparison Between Physico-chemical Expected and Measured Adhesion of Oxygen-plasma-treated Carbon Fibres and Polycarbonate. *J. Adhes.* **2000**, *73*, 19–42. [CrossRef]

10. Bismarck, A.; Kumru, M.E.; Song, B.; Springer, J.; Moos, E.; Karger-Kocsis, J. Study on surface and mechanical fibre characteristics and their effect on the adhesion properties to a polycarbonate matrix tuned by anodic carbon fibre oxidation. *Compos. Part A Appl. Sci. Manuf.* **1999**, *30*, 1351–1366. [CrossRef]

11. Yao, T.T.; Wu, G.P.; Song, C. Interfacial adhesion properties of carbon fibre/polycarbonate composites by using a single-filament fragmentation test. *Compos. Sci. Technol.* **2017**, *149*, 108–115. [CrossRef]

12. Lee, J.; Drzal, L.T. Surface characterization and adhesion of carbon fibres to epoxy and polycarbonate. *Int. J. Adhes.* **2005**, *25*, 389–394. [CrossRef]

13. Raghavendran, V.K.; Drzal, L.T.; Askeland, P. Effect of surface oxygen content and roughness on interfacial adhesion in carbon fibre–polycarbonate composites. *J. Adhes. Sci. Technol.* **2002**, *16*, 1283–1306. [CrossRef]

14. Lorca, J.L.; Gonzalez, C.; Molina-Aldarequia, J.M.; Segurado, J.; Seltzer, R.; Sket, F.; Rodriguez, M.; Sadaba, S.; Munoz, R.; Canal, L.P. Multiscale Modeling of composite materials: A Roadmap towards virtual testing. *Adv. Mater.* **2011**, *23*, 5130–5147. [CrossRef] [PubMed]

15. Pisanova, E.V.; Zhandarov, S.F.; Dovgyalo, V.A. Interfacial adhesion and failure modes in single filament thermoplastic composites. *Polym. Compos.* **1994**, *15*, 147–155. [CrossRef]

16. Mäder, E.; Pisanova, E. Interfacial design in fibre reinforced polymers. *Macromol. Symp.* **2001**, *163*, 189–212. [CrossRef]

17. Kim, J.-K.; Mai, Y.-M. *Engineered Interfaces in Fibre Reinforced Composites*; Elsevier: Amsterdam, The Netherlands, 1998.

18. Zhandarov, S.; Mäder, E. Characterization of fibre/matrix interface strength: Applicability of different tests, approaches and parameters. *Compos. Sci. Technol.* **2005**, *65*, 149–160. [CrossRef]

19. Kamps, J.H.; Scheffler, C.; Simon, F.; van der Heijden, R.; Verghese, N. Functional polycarbonates for improved adhesion to carbon fibre. *Compos. Sci. Technol.* **2018**, *167*, 448–455. [CrossRef]

20. Jones, M.D.; Hooton, J.C.; Dawson, M.L.; Ferrie, A.R.; Price, R. An investigation into the dispersion mechanisms of ternary dry powder inhaler formulations by the quantification of interparticulate forces. *Pharm. Res.* **2008**, *25*, 337–348. [CrossRef] [PubMed]

21. Good, R.J.; Srivatsa, N.R.; Islam, M.; Huang, H.T.L.; van Oss, C.J. Theory of the acid-base hydrogen bonding interactions, contact angles, and the hysteresis of wetting: Application to coal and graphite surfaces. *J. Adhes. Sci. Technol.* **1990**, *4*. [CrossRef]

22. Rulison, C. *So You Want to Measure Surface Energy?—A Tutorial Designed to Provide Basic Understanding of the Concept of Solid Surface Energy, and Its Many Complications*; Krüss Technical Note #306; KRÜSS GmbH: Hamburg, Germany, 1999.

23. Mäder, E.; Grundke, K.; Jacobasch, H.J.; Wachinger, G. Surface, interphase and composite property relation in fibre-reinforced polymers. *Composites* **1994**, *25*, 739–744. [CrossRef]

24. Miller, B.; Muri, P.; Rebenfeld, L. A microbond method for determination of the shear strength of a fibre–resin interface. *Compos. Sci. Technol.* **1987**, *28*, 17–32. [CrossRef]

25. Zhandarov, S.; Mäder, E. An alternative method of determining the local interfacial shear strength from force-displacement curves in the pull-out and microbond tests. *Int. J. Adhes.* **2014**, *55*, 37–42. [CrossRef]

26. Donnet, J.-B.; Bansal, R.C. *Carbon Fibers*, 2nd ed.; Marcel Dekker: New York, NY, USA; Basel, Switzerland, 1990; pp. 17–32.

27. Tripathi, B.; Das, P.; Simon, F.; Stamm, M. Ultralow fouling membranes by surface modification with functional polydopamine. *Eur. Polym. J.* **2018**, *99*, 80–89. [CrossRef]

28. Beamson, G.; Briggs, D. High resolution of organic polymers. In *The Scienta ESCA 300 Database*; J. Wiley & Sons: Chichester, NY, USA; Brisbane: Toronto, ON, Canada; Singapore, 1992; pp. 184–187; ISBN 0-471-93592-1.

29. Rulison, C. *Adhesion Energy and Interfacial Tension—Two Related Coating/Substrate Interfacial Properties—Which Is More Important for Your Application, and Why?* Krüss Application Note #232e; KRÜSS GmbH: Hamburg, Germany, 2003.

materials

MDPI

Article

Curing Effects on Interfacial Adhesion between Recycled Carbon Fiber and Epoxy Resin Heated by Microwave Irradiation

Yuichi Tominaga *, Daisuke Shimamoto and Yuji Hotta

National Institute of Advanced Industrial Science and Technology (AIST), Shimosidami, Moriyama-ku, Nagoya 463-8560, Japan; d-shimamoto@aist.go.jp (D.S.); y-hotta@aist.go.jp (Y.H.)
* Correspondence: tominaga.yuichi@aist.go.jp; Tel.: +81-52-736-7498

Received: 8 March 2018; Accepted: 23 March 2018; Published: 26 March 2018

Abstract: The interfacial adhesion of recycled carbon fiber (CF) reinforced epoxy composite heated by microwave (MW) irradiation were investigated by changing the curing state of the epoxy resin. The recycled CF was recovered from the composite, which was prepared by vacuum-assisted resin transfer molding, by thermal degradation at 500 or 600 °C. Thermogravimetric analysis showed that the heating at 600 °C caused rough damage to the CF surface, whereas recycled CF recovered at 500 °C have few defects. The interfacial shear strength (IFSS) between recycled CF and epoxy resin was measured by a single-fiber fragmentation test. The test specimen was heated by MW after mixing the epoxy resin with a curing agent or pre-curing, in order to investigate the curing effects on the matrix resin. The IFSSs of the MW-irradiated samples were significantly varied by the curing state of the epoxy resin and the surface condition of recycled CF, resulting that they were 99.5 to 131.7% of oven heated samples Furthermore, rheological measurements showed that the viscosity and shrinking behaviors of epoxy resin were affected based on the curing state of epoxy resin before MW irradiation.

Keywords: interfacial adhesion; recycled carbon fiber; microwave heating; epoxy curing

---

1. Introduction

Carbon fiber-reinforced plastic (CFRP) is attracting attention due to its excellent mechanical and light-weight properties [1–7]. Because the production of CFRPs has increased, there is increasing environmental and economic awareness for the need to recycle CFRP waste. Typically, CFRP waste includes various CFRPs, such as expired prepregs, manufacturing cut-offs, testing materials, production tools, and end-of-life components. Furthermore, recycled CFs can be recovered from CFRP wastes by various recycling processes, such as thermal degradation [8,9], chemical decomposition [10,11], and superheated steam [12], due to which the recycled CF has some surface defects, such as pitting, residual matrix, and char. In addition, all recycling processes remove the sizing agent from CF surfaces; hence, the interfacial adhesion between recycled CF and matrix resin is reduced. In the case of commercial virgin CF, the surface of CF is coated with sizing agents for the improvement of interfacial adhesion by a sizing apparatus for continuous fiber. However, it is difficult to apply the manufacturing processes developed for virgin continuous CF, because recycled CF is discontinuous. A complicated process to improve interfacial adhesion is not practical from the viewpoint of manufacturing cost. Therefore, it is critical to develop a simple process to improve interfacial adhesion between various recycled CFs and resin.

Recently, microwave (MW) processes have attracted attention as an innovative curing system for composites composed of CF and thermosetting resins [13,14]. In previous study, the interfacial adhesion between virgin CF and thermosetting epoxy resin was enhanced by MW heating [15]. In this process,

CFRP is heated from the inside by the selective heating of CF in the composite. The chemical reaction of epoxy resin and curing agent occurs preferentially on the surface of CF, suggesting that preferential reactions of epoxy resin on the CF interface lead to the suppression of interface delamination, thereby strengthening interfacial adhesion.

The main objective of this study was to clarify the effects of the curing state of epoxy resin and the surface condition of recycle CF on interfacial adhesion between various recycled CFs and epoxy resin as the thermosetting resin matrix heated by MW irradiation. Furthermore, the viscosity and the shrinkage behavior of epoxy resin at the curing temperature were investigated by rheological measurements.

2. Materials and Methods

Commercially available CF fabrics (Torayca BT70-20, Toray, Japan) were used in this work. The textile weight of used CF fabrics is 100 g/m^2. Bisphenol A-type epoxy resin (JER827), bisphenol F-type epoxy resin (JER806), and curing agents (JER cure 3080 or ST11) were purchased from Mitsubishi Chemical Co. Ltd., Tokyo, Japan. These epoxy materials are listed in Table 1. Epoxy resin JER806 was mixed with curing agent ST11 at a mass ratio of 5:3 using a hybrid deformation blender (ARE-310, Thinky, Tokyo, Japan) for 1 min at 2000 rpm. The epoxy resin was penetrated by vacuum-assisted resin transfer molding (VaRTM) to CF fabrics consisting of four plies. The CFRP was pre-cured by maintaining the sample at ambient temperature for 24 h. Post-curing of the composite was performed at 120 °C for 3 h by muffle furnace with an exhaust port (NMF-215B, Masuda Corporation, Chiba, Japan). The volume fiber fraction (V$_f$) of the prepared CFRP was measured using JIS K7075 [16]. The V$_f$ was 65 vol %. The recycled CF was recovered from prepared CFRP by thermal degradation processing at 500 or 600 °C for 1 h in an air atmosphere. In the recycling process, the heating rate was adjusted to 5 °C/min. Hereafter, virgin CF, recycled CF heated at 500 °C, and recycled CF heated at 600 °C are termed vCF, rCF-500, and rCF-600, respectively.

Table 1. Materials for composites.

Sample	Epoxy		Curing Agent		Mass Ratio of Epoxy to Curing Agent (w/w)
	Type	EEW [1] (g/eq)	Type	AEW [2] (KOHmg/g)	
CFRP for recovering recycled CF	JER806 (Bisphenol F)	160–170	ST11 (Amine)	325–360	5/3
Composite for fragmentation test	JER827 (Bisphenol A)	180–190	3080 (Amine)	310–340	2/1

[1] Epoxy equivalent weight; [2] Amine equivalent weight.

Thermogravimetric (TG) analyses of CF and CFRP were performed using a thermal analysis instrument (TG-8120, Rigaku, Tokyo, Japan) from 30 °C to 900 °C under natural air of 200 mL/min at a heating rate of 5 °C/min. Furthermore, TG analyses of CFRP were carried out at temperatures of 500 and 600 °C for 3 h at a heating rate of 5 °C/min.

Tensile testing of a single CF was performed using an autograph apparatus (EZ-SX, Shimadzu, Kyoto, Japan) with a load cell of 10 N. Test specimens for tensile testing were prepared by fixing a single CF onto a paper with an adhesive agent. The test specimen was set in the tensile apparatus and the paper was cut into two parts before evaluation. Single-fiber tensile tests were carried out at a gauge length of 25 mm. The cross-head speed was 1.0 mm/min. The tensile strength of the CFs was calculated using the standard Weibull distribution using Equations (1) and (2) [17,18]. The average diameters of the CFs were estimated by field emission scanning electron microscopy (FE-SEM, S-4300, Hitachi, Tokyo, Japan).

$$F(\sigma) = 1 - P_0 = 1 - exp(-(\frac{\sigma}{\gamma})^\beta) \tag{1}$$

$$\sigma = \gamma \Gamma(1 + \frac{1}{\beta}) \tag{2}$$

where β is the shape parameter, γ is the scale parameter, $\Gamma(x)$ is the gamma function, and σ is the tensile strength.

A single CF was placed in the center of a dog-bone shaped silicon mold. The gauge length of the test specimens for fragmentation testing was fixed at 35 mm. Each end of the CF was fixed with adhesive tape. Epoxy resin JER827 was mixed with curing agent JER cure 3080 at a mass ratio of 2:1 using a hybrid deformation blender for 1 min at 2000 rpm. Epoxy resin was cast into the silicon mold. For this test, the test specimen was irradiated by MW at different times. One was MW-irradiated immediately after resin mixing, and the sample was pre-cured at 23 °C for 24 h. Then, the sample was post-cured at 80 °C for 3 h. The other sample was MW-irradiated after pre-curing at 23 °C for 24 h, and the specimen was post-cured. In this examination, the absorption of MW irradiation of the specimen could not be controlled, as the sample was a single CF. Therefore, the output powers of MW irradiation were controlled at 10, 20, and 30 W. The MW irradiation absorbed by the specimen was calculated by subtracting the reflection power from the output power.

The IFSS between CF and epoxy resin was immediately measured without any conditioning after the preparation of test specimen by a single-fiber fragmentation test. The fragmentation test of the prepared composite was performed using a compact tensile apparatus at a displacement rate of 10 mm/s. The specimen was pulled until the fragments of fiber no more break. During this test, the specimen was observed using a microscope (BX-51, Olympus, Tokyo, Japan) with a digital camera (DP21, Olympus, Tokyo, Japan). The fragment length within the gauge length was measured via image analysis with a calibrated measuring scale. The IFSS between the carbon fiber and epoxy resin was calculated using the following Equation (3) [19–22]:

$$\tau_i = \frac{D\sigma_f}{2l_c} \tag{3}$$

where σ_f is tensile strength of CF, D is diameter of CF, and l_c is the critical fiber length. The critical fiber length (l_c) can be determined using the average fragment length (l) by Equation (4):

$$l_c = \frac{4}{3}l \tag{4}$$

The average fragment length is the arithmetic mean of the measured fragment lengths at the saturation stage [20]. The IFSS was the average of seven measurements.

The shrinkage and rheological behavior of the epoxy resin composed of JER827 and JER cure 3080 were measured using a rheometer (MCR102, Anton Paar, Graz, Austria) Disposable aluminum parallel plates with diameters of 25 mm were used in the oscillation mode. The normal force is the amount of force applied by the top plate and is defined as negative when the epoxy resin is in compression. The test was performed in the torque- or strain-controlled mode at a constant angular velocity. Immediately after mixing, the epoxy resin was poured between the plates at room temperature and the gap was set at 1.0 mm. After pouring, the epoxy resin was maintained at 25 °C for 5 min, and heated to 80, 100, and 120 °C until its gelation point was reached. The gelation point of epoxy resin was determined based on the cross-over point of the elastic modulus and loss modulus. After gelation, the sample was heated further at the same temperature for 1 h and post-heated at 120 °C for 1 h. Finally, the sample was cooled to 25 °C. The rheometer was programmed with two segments. In segment I, the gap between the plates was kept constant and the strain was fixed at 15% with an oscillation frequency of 0.2 Hz. When the epoxy resin reached the gelation point, the rheometer was controlled according to segment II. In segment II, the torque was fixed at 0.5 mN with an oscillation frequency of 30 Hz and the normal force was kept constant at 0.1 N. The typical temperature profile, gap, and normal force of epoxy resin at a heating temperature of 100 °C are shown in Figure 1. The rheological behavior of the epoxy resin heated after pre-curing at room temperature was also measured according to the following conditions. The epoxy resin was pre-cured at 23 and 30 °C for 24 h. After pre-curing, the sample was post-heated at 120 °C for 3 h and cooled to 25 °C. The rheometer was also programmed with two segments. In segment a, the gap between the plates was kept constant and the strain was fixed at 1.0% with an oscillation frequency of 1.0 Hz. When the epoxy resin reached the gelation point, the rheometer was controlled by segment b, in which the strain was fixed at 0.02% with an oscillation frequency of 1.0 Hz and a constant normal force of 0.1 N. The linear cure shrinkage ε_L was calculated using the following the Equation (5) [23]:

$$\varepsilon_L = \left[(1+\nu)\left(\frac{\nu}{1-\nu}+1\right)\right]^{-1}(\frac{h_0-h}{h_0}) \tag{5}$$

where ν is Poisson's coefficient of the resin, taken to be 0.35 [24,25], h is the gap distance, and h_0 represents the initial gap. Here, this relation assumes that the in-plane strain of the resin is zero.

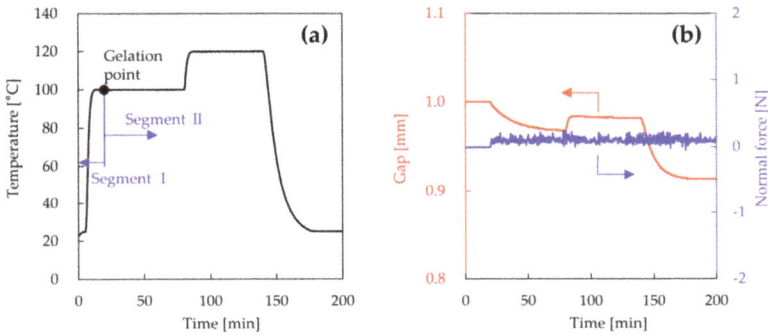

Figure 1. Temperature profile (**a**); gap and normal force (**b**) of epoxy resin at 100 °C setting temperature.

3. Results and Discussion

Figure 2a shows the TG curves of CF and CFRP as a function of temperature. The weight of CFRP decreased from about 300 °C and decreased greatly at approximately 400 °C. The weight corresponding to the epoxy resin decreased at 300–550 °C. The weight of CFRP decreased further from 600 °C, and the CFRP burned completely at 830 °C. The weight of virgin CF decreased slowly until 600 °C, and virgin CF was destroyed between 600 °C and 800 °C. These results suggest that recycled CF can be recovered from CFRP by thermal degradation at 600 °C or less. Figure 2b shows the TG curves of CFRP as a function of time under isothermal conditions of 500 and 600 °C. The weight losses of CFRP at 500 °C for 0, 1, 2, and 3 h were 26.0, 30.5, 31.6, and 33.1%, respectively. The weight of CFRP decreased slightly by maintaining the temperature at 500 °C. On the other hand, the weight losses of CFRP at 600 °C for 0, 1, 2, and 3 h were 29.4, 46.9, 64.9, and 80.5%, respectively. The weight of CFRP decreased greatly with increasing heating time, indicating that the CF underwent severe decomposition at 600 °C under air. In this study, to investigate the influence of recycled CF with different surface conditions, the temperatures for the thermal degradation recycling process of CFRP were set to 500 and 600 °C for 1 h. After the thermal degradation of bulk CFRP for 1 h in a muffle furnace, the weight losses of CFRP heated at 500 and 600 °C were 32.8 and 45.4%, respectively. These weight losses agreed with the results of TG measurements. The recovery rates of recycled CF heated at 500 and 600 °C estimated from the weight loss were 95.4 and 68.9%, respectively. Therefore, the epoxy resin of CFRP have been completely pyrolyzed in this thermal degradation process.

Figure 3 shows the FE-SEM images of vCF, rCF-500, and rCF-600. The surface morphology of rCF-500 was almost identical to vCF. On the other hand, the surfaces of several rCF-600 were obviously degraded. The diameters of vCF, rCF-500, and rCF-600 measured from the FE-SEM images were 7.24, 6.87 and 6.21 mm, respectively. The diameter of recycled CF decreased with increasing heating temperature. Figure 4 shows the Weibull plot of the tensile strength of CF. The plot of rCF-600 was not linear and did not conform to the Weibull plot, although the plots of vCF and rCF-500 were almost linear. As shown in Figure 3, rCF-600 contains damaged and undamaged CFs. Because various kinds of defects are introduced into the CFs due to the decomposition of CF, the destruction mode of rCF-600 varies depending on the kinds of the surface defect. Therefore, the rCF-600 did not to conform to the Weibull plot. From these results, the arithmetic mean of the tensile strength of CF with a gauge length of 25 mm was used as the tensile strength of CF. The tensile strengths of vCF, rCF-500, and rCF-600 were 3.60, 3.38, and 2.75 GPa, respectively. The tensile strength of recycled CF decreased with increasing heating temperature.

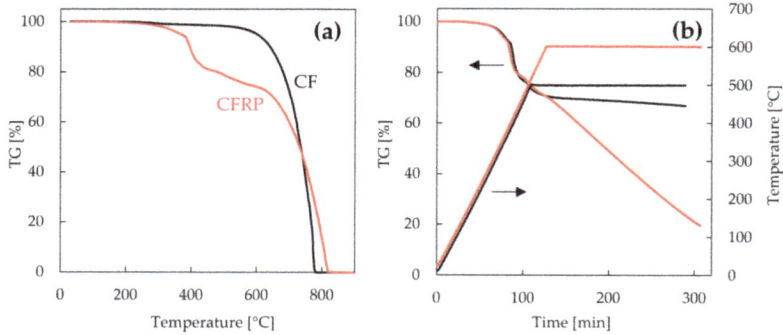

Figure 2. Thermogravimetric (TG) curves of carbon fiber (CF) and carbon fiber-reinforced plastic (CFRP) as a function of temperature (**a**) and CFRP as a function of time under isothermal conditions of 500 and 600 °C (**b**).

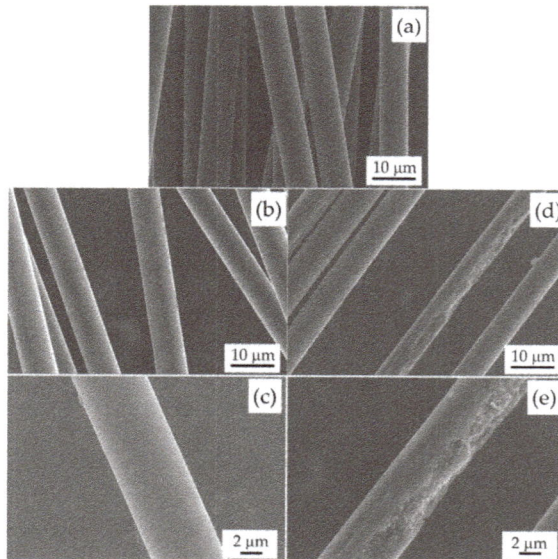

Figure 3. FE-SEM images of vCF (**a**); rCF-500 (**b**,**c**) and rCF-600 (**d**,**e**).

Table 2 shows the IFSS of the samples cured in a conventional oven. Generally, IFSS is calculated by substituting the fiber ultimate tensile strength at the broken fiber length, which is calculated using a simple weakest-link scaling, into Equation (3) [26,27]. However, rCF-600 did not conform to the Weibull distribution. Therefore, in this study, in order to compare different recycled CFs, the IFSS was calculated by substituting the arithmetic mean of tensile strength of CF with a gauge length of 25 mm into Equation (3). The IFSSs of oven-cured samples composed of vCF, rCF-500, and rCF-600 were 23.8, 18.0, and 14.4 MPa, respectively. The IFSSs of samples prepared with recycled CFs were lower compared with vCF, indicating that the IFSS was decreased by removal of sizing agent in recycling process. Furthermore, the IFSS of the rCF-600 sample was lower compared with rCF-500. This result suggests that the epoxy resin does not impregnate the surface of rCF-600 due to the large defects on the surface.

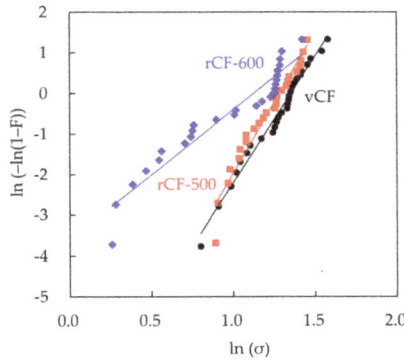

Figure 4. Weibull plots of tensile strengths of vCF (black), rCF-500 (red), and rCF-600 (blue).

Table 2. Interfacial shear strength (IFSS) of the samples cured in a conventional oven.

CF	Critical Fiber Length, l_c (mm)	IFSS [1] (MPa)	IFSS [2] (MPa)
vCF	550 (±34)	46.0 (±3.4)	23.8 (±1.5)
rCF-500	645 (±27)	31.8 (±1.5)	18.0 (±0.8)
rCF-600	597 (±35)	-	14.4 (±0.9)

[1] IFSS calculated from the fiber ultimate tensile strength at the broken fiber length; [2] IFSS calculated from the arithmetic mean of fiber strength of CF with a gauge length of 25 mm.

Figure 5 shows the IFSS of MW-irradiated samples prepared with recycled CF as a function of absorbed MW. Here, the sample was MW-irradiated for 5 min after epoxy resin penetrated the CF and then cured in a conventional oven. In MW irradiation, CF is heated selectively. Because the thermal conductivity of epoxy resin is low, the epoxy resin is cured only at the interface with CF, and the heat of a single CF is not sufficiently transferred to the epoxy resin. Unfortunately, the long duration of MW irradiation results in the thermal degradation of epoxy resin at the interface with CF [15]. Thus, the specimen for the fragmentation test is not cured completely by MW irradiation. In order to evaluate quantitatively the interfacial adhesion between CF and epoxy resin by the fragmentation test, the epoxy resin sample must be cured completely. Therefore, the sample which was cured on the CF interface by MW irradiation was further cured in a conventional oven. The IFSSs of rCF-500 specimens that absorbed MW of 0, 1.3, 2.6, and 4.7 W were 18.0, 20.3, 21.9, and 22.9 MPa, respectively. Additionally, the IFSSs of rCF-600 specimens that absorbed MW of 0, 1.2, 2.6, and 4.0 W were 14.4, 16.6, 16.9, and 17.6 MPa, respectively. The IFSSs of MW-assisted recycled CF samples increased as compared to those without MW irradiation. Furthermore, the IFSSs of samples increased with increasing amounts of absorbed MW. Figure 6 shows the IFSSs of samples prepared by varying the MW irradiation time and the curing state of epoxy resin. The IFSSs of samples MW-irradiated for 10 min were similar to those MW-irradiated for 5 min, indicating that the state of the interface of MW-irradiated samples for 10 min was comparable to those irradiated for 5 min. This result suggests that the epoxy resin on the interface of CF is cured in a short time (within 5 min). The IFSSs of rCF-500 samples MW-irradiated after pre-curing were similar to those MW-irradiated after penetrating. On the other hand, in the case of rCF-600, IFSSs decreased by MW-irradiation after pre-curing. The major difference between rCF-500 and the rCF-600 is the roughness of the surface of recycled CF, suggesting that the roughness of the surface of CF and the curing state of epoxy resin affect interfacial adhesion.

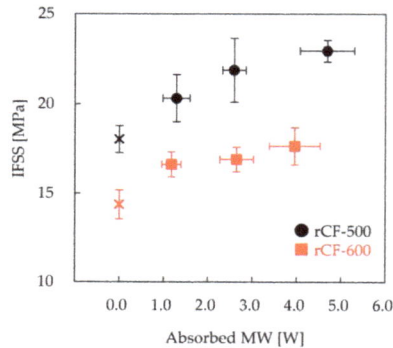

Figure 5. IFSSs of MW-irradiated samples prepared with rCF-500 (black symbol) and rCF-600 (red symbol) as a function of absorbed MW.

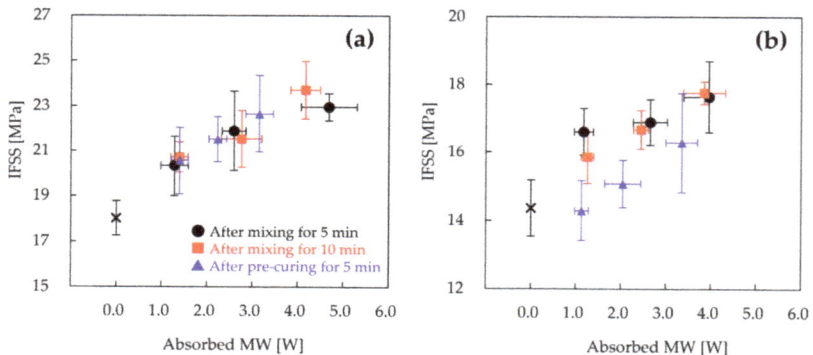

Figure 6. IFSSs of samples prepared with rCF-500 (**a**) and rCF-600 (**b**). The samples were MW-irradiated for 5 min (black circle) or 10 min (red square) after mixing the epoxy resin and curing agent. The sample was MW-irradiated after pre-curing at 23 °C for 24 h (blue triangle).

The improvement of the interfacial adhesion between recycled CF and epoxy resin by MW irradiation is now discussed. The improved IFSS is assumed to be mainly derived from the viscosity and shrinkage of the epoxy resin. For the purpose of validating whether the viscosity and the shrinkage of epoxy resin influence the interfacial adhesion, the viscosity variation and the shrinkage behavior of epoxy resin were investigated by rheological measurements. In previous studies, the temperature of CF increased with increasing output power of MW [15]. In fact, the recycled CF was burned off at high output powers of 40 W or more. Because it was very difficult to measure the temperature at the interface during MW irradiation, the viscosity and shrinkage of epoxy resin were measured by changing the heating temperature. Previous work suggests that the resin-fiber interface and the fiber wet-out can be improved by lowering the resin viscosity [14]. Figure 7a shows the complex viscosity of epoxy resin heated at different temperatures after mixing as a function of time. The viscosity of the used epoxy resin decreased considerably by heating at high temperature, following which the epoxy resin became very viscous due to crosslinking. The viscosities of the resins heated at 100 and 120 °C were reduced greatly from 3500 MPa·s to 40 MPa·s, and were lower than that of the resin heated at 80 °C. Furthermore, the viscosity of epoxy resin decreases as the heating rate increases [14]. On the other hand, the viscosities of samples heated after pre-curing varied greatly depending on the temperature of pre-curing, as shown in Figure 7b. The viscosities of the resin heated at 120 °C after pre-curing at 30 and 23 °C were 1.19×10^8 and 2.10×10^4 MPa·s, respectively. The viscosity of the resin pre-heated at 23 °C, which was not gelled for 24 h, decreased greatly. However, the viscosity of the resin pre-heated at 30 °C, which was gelled for 16.5 h, only minimally decreased. Although the viscosity of the resin heated after pre-curing at 23 °C was reduced by heating, the viscosity remained higher than the initial viscosity of the resin. In MW heating, CFs act as a conductor and get abruptly heated due to ionic polarization and dipole orientation, rather than relying on heat transfer through conduction and convection [14,28–30]. Therefore, the temperature of the resin at the interface should increase rapidly. The viscosity reduction of the epoxy resin at the interface, based on the high heating rate, would make it easier for the epoxy resin to impregnate the surface of CF in the initial stage of the curing process. Thus, excellent impregnation behavior of epoxy resin by MW irradiation leads to an increase of the reaction efficiency of CF and epoxy resin. In the case of rCF-600, the epoxy resin would be easily impregnated into the rough surface of rCF-600 by irradiating with MW immediately after mixing. On the other hand, MW irradiation after pre-curing might suppress the impregnation of the epoxy resin to the surface of rCF-600. Therefore, the result suggests that the IFSS of rCF-600 samples MW-irradiated after pre-curing would drop due to the lowering of the impregnation based on the high viscosity of the epoxy resin. In the case of the rCF-500, due to the smooth surface of rCF-500, resin impregnation would be unaffected by interfacial adhesion, resulting in IFSSs of MW-irradiated samples after pre-curing to be similar to those irradiated after mixing.

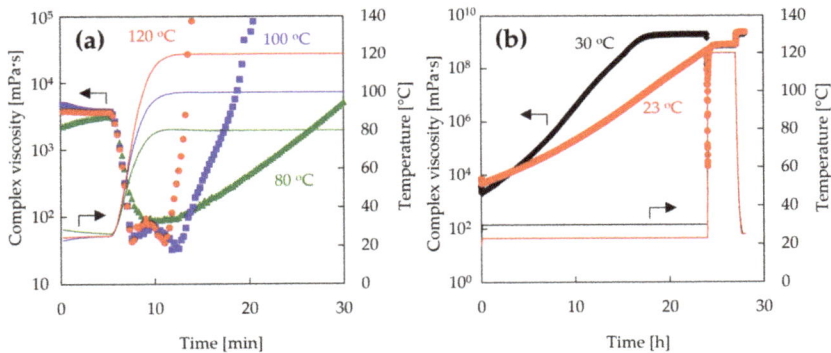

Figure 7. Complex viscosity of epoxy resin at different temperatures as a function of time. The epoxy resin was heated after mixing (**a**) or pre-curing (**b**).

The effect of shrinkage of the epoxy resin on interfacial adhesion is also discussed. Figure 8a shows the linear cure shrinkage of epoxy resin after mixing as a function of time. For a curing temperature of 100 °C, the epoxy resin shrunk by curing and then thermally expanded by heating at 120 °C. Finally, the cured epoxy resin thermally shrunk significantly when returning to room temperature. This shrinking behavior of epoxy resin after mixing was observed at all setting temperatures. The linear cure shrinkages of the epoxy resin heated at 80, 100, and 120 °C from the initial stage to when returning to room temperature were 2.65, 4.14, and 4.96%, respectively. The linear cure shrinkage of the epoxy resin increased with increasing initial curing temperature. Figure 8b also shows the linear cure shrinkage of epoxy resin after pre-curing, as a function of time. Furthermore, the linear cure shrinkages of the epoxy resin heated at 120 °C after pre-curing at 23 and 30 °C were 4.62 and 1.01%, respectively. The linear cure shrinkage of the gelled epoxy resin was small even when it was cured at 120 °C. On the other hand, the linear cure shrinkages of the non-gelled samples at 23 °C were almost the same as that of the sample heated at 120 °C immediately after mixing. The pre-curing of epoxy resin before gelation had no effect on the shrinking behavior of the epoxy resin. This result indicates that the temperature during gelation of the epoxy resin is important for the shrinkage of the epoxy resin. These results also showed that the linear cure shrinkage of the sample MW-irradiated after pre-curing were similar to the one MW-irradiated after penetrating. In MW heating, the temperature of the epoxy resin on the CF interface rises rapidly. Furthermore, it is reported that the polarity functional groups with an active oxygen in CF are locally-heated strongly rather than the non-polarity functional groups [31,32]. Therefore, the polarity functional groups can react with the epoxy groups in the epoxy resin and the epoxy resin on the CF interface binds tightly to the CF surface [33,34]. Moreover, in a previous study, the activation energy of the CFRP prepared by MW heating was about one half as compared to the one prepared by conventional heating [35]. Because the reaction rate of epoxy resin is increased by lowering the activation energy, the interfacial adhesion is also improved. Furthermore, the heat generated from CF is diffused outward from the CF, resulting in curing of the epoxy resin which should progress outward from the CF. Figure 9 shows the SEM image of recycled CF, which was MW-irradiated for 5 min and washed with ethanol to remove unreacted epoxy resin. It was confirmed that the epoxy resin was adsorbed on the surface of CF, showing that the epoxy resin was cured from the CF interface. Furthermore, in the MW process, the curing of the epoxy resin outward from the CF interface causes the shrinkage of the epoxy resin along the CF direction. Because the temperature of the epoxy resin near the fiber is high, the shrinkage of the epoxy resin at the interface is significant. As results, the epoxy resin can hug the CF strongly. However, in the case of conventional oven heating, the epoxy resin in the composite is heated from the outside, due to which the chemical reaction of the epoxy resin is initiated from the outside. Because the curing reaction is an exothermic reaction, the reaction of the epoxy resin from the outside is accelerated, and the epoxy resin shrinks from the outside. This shrinking behavior of epoxy resin from the outside causes the delamination of the epoxy resin from CF surface. In fact, the interface of the composite of CF and epoxy resin cured by conventional oven was partially delaminated [15]. Furthermore, the reactivity of the epoxy resin with CF must be lesser than with MW heating, because the temperature of epoxy resin on the CF interface is lower than the setting temperature. Thus, in the conventional oven heating, the interfacial adhesion would decrease. These results suggest that external heating, such as that in an electric oven, requires slow curing at a low temperature for the suppression of shrinkage, whereas MW irradiation can facilitate rapid curing with high-temperature heating. MW irradiation can thus improve the interfacial adhesion between recycled CF and thermosetting resins via mainly shrinkage of epoxy resin along the fiber direction and enhanced impregnation owing to the decrease of viscosity of the resin on the interface.

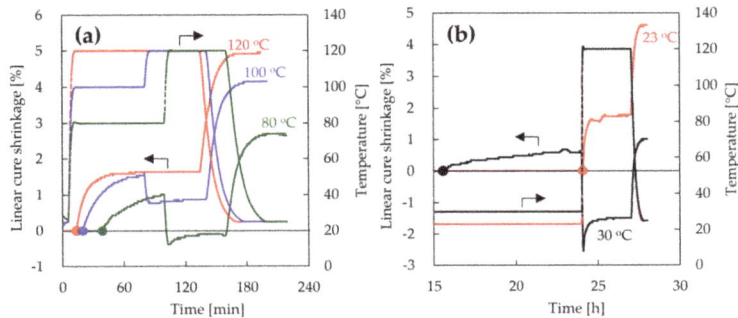

Figure 8. Linear cure shrinkage of epoxy resin as a function of time. Epoxy resin was heated after mixing (**a**) or pre-curing (**b**). The circle symbol is the gelation point of epoxy resin.

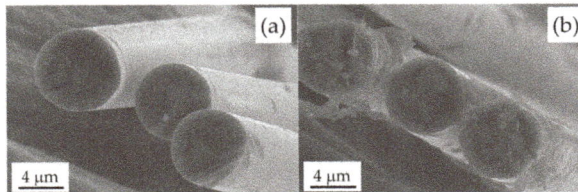

Figure 9. FE-SEM images of rCF-500 (**a**) and rCF-500 (**b**), which were MW-irradiated for 5 min and washed with ethanol to remove unreacted epoxy resin.

4. Conclusions

To improve the interfacial adhesion between various recycled CFs and epoxy resin by MW irradiation, the IFSSs of composites prepared under different conditions was investigated. In this study, recycled CFs, with or without surface defects, were recovered from CFRP by thermal degradation processes at different temperatures. In the case of the composite MW-irradiated immediately after mixing, the IFSSs of both recycled CFs samples increased with increasing absorbed MW. On the other hand, the IFSS of the composite composed of recycled CF with surface defects did not improve by MW irradiation after pre-curing. We also investigated the curing behavior of epoxy resins by using a rheometer. The viscosity of the epoxy resin decreased greatly by heating immediately after mixing, whereas its reduction was suppressed by heating after pre-curing. The pre-curing of epoxy resin before gelation did not affect the shrinking behavior of epoxy resin. These results suggest that the MW process improved the interfacial adhesion between recycled CF and epoxy resin from two aspects, namely, by enhanced impregnation owing to the decrease of viscosity of the resin on the interface, and the shrinkage of epoxy resin along the fiber direction.

One important consequence of the present work is that the interfacial adhesion between recycled CF and the epoxy resin is significantly affected by the curing state of the epoxy resin before MW irradiation. The interfacial adhesion between various recycled CF and epoxy resin can be improved by MW irradiation immediately after mixing the epoxy resin and curing agent.

Acknowledgments: We express our sincere thanks to Shunichi Furusho of Anton Paar Japan Co., Ltd. for supporting this study. We thank Yusuke Imai, Kimiyasu Sato, and Yoshiki Sugimoto for their comments.

Author Contributions: Yuichi Tominaga, Daisuke Shimamoto and Yuji Hotta conceived and designed the experiments; Yuichi Tominaga performed the experiments; all authors analyzed and discussed the experimental results; and Yuichi Tominaga wrote the paper.

Conflicts of Interest: The authors declare no conflict of interest.

References

1. Karnik, S.; Gaitonde, V.; Rubio, J.C.; Correia, A.E.; Abrão, A.; Davim, J.P. Delamination analysis in high speed drilling of carbon fiber reinforced plastics (CFRP) using artificial neural network model. *Mater. Des.* **2008**, *29*, 1768–1776. [CrossRef]
2. Rezaei, F.; Yunus, R.; Ibrahim, N.A. Effect of fiber length on thermomechanical properties of short carbon fiber reinforced polypropylene composites. *Mater. Des.* **2009**, *30*, 260–263. [CrossRef]
3. Ashrafi, B.; Guan, J.; Mirjalili, V.; Zhang, Y.; Chun, L.; Hubert, P.; Simard, B.; Kingston, C.T.; Bourne, O.; Johnston, A. Enhancement of mechanical performance of epoxy/carbon fiber laminate composites using single-walled carbon nanotubes. *Compos. Sci. Technol.* **2011**, *71*, 1569–1578. [CrossRef]
4. Davis, D.C.; Wilkerson, J.W.; Zhu, J.; Hadjiev, V.G. A strategy for improving mechanical properties of a fiber reinforced epoxy composite using functionalized carbon nanotubes. *Compos. Sci. Technol.* **2011**, *71*, 1089–1097. [CrossRef]
5. Khan, S.U.; Kim, J.K. Improved interlaminar shear properties of multiscale carbon fiber composites with bucky paper interleaves made from carbon nanofibers. *Carbon* **2012**, *50*, 5265–5277. [CrossRef]
6. Alizadeh Ashrafi, S.; Miller, P.W.; Wandro, K.M.; Kim, D. Characterization and effects of fiber pull-outs in hole quality of carbon fiber reinforced plastics composite. *Materials* **2016**, *9*, 828. [CrossRef] [PubMed]
7. Zhang, K.; Tang, W.; Fu, K. Modeling of dynamic behavior of carbon fiber-reinforced polymer (CFRP) composite under X-ray radiation. *Materials* **2018**, *11*, 143. [CrossRef] [PubMed]
8. Marsh, G. Reclaiming value from post-use carbon composite. *Reinf. Plast.* **2008**, *52*, 36–39. [CrossRef]
9. Meyer, L.; Schulte, K.; Grove-Nielsen, E. Cfrp-recycling following a pyrolysis route: Process optimization and potentials. *J. Compos. Mater.* **2009**, *43*, 1121–1132. [CrossRef]
10. Jiang, G.; Pickering, S.J.; Lester, E.H.; Turner, T.; Wong, K.; Warrior, N. Characterisation of carbon fibres recycled from carbon fibre/epoxy resin composites using supercritical n-propanol. *Compos. Sci. Technol.* **2009**, *69*, 192–198. [CrossRef]
11. Piñero-Hernanz, R.; Dodds, C.; Hyde, J.; García-Serna, J.; Poliakoff, M.; Lester, E.; Cocero, M.J.; Kingman, S.; Pickering, S.; Wong, K.H. Chemical recycling of carbon fibre reinforced composites in nearcritical and supercritical water. *Compos. Part A Appl. Sci. Manuf.* **2008**, *39*, 454–461. [CrossRef]
12. Shi, J.; Bao, L.; Kobayashi, R.; Kato, J.; Kemmochi, K. Reusing recycled fibers in high-value fiber-reinforced polymer composites: Improving bending strength by surface cleaning. *Compos. Sci. Technol.* **2012**, *72*, 1298–1303. [CrossRef]
13. Brosseau, C.; Quéffélec, P.; Talbot, P. Microwave characterization of filled polymers. *J. Appl. Phys.* **2001**, *89*, 4532–4540. [CrossRef]
14. Papargyris, D.A.; Day, R.J.; Nesbitt, A.; Bakavos, D. Comparison of the mechanical and physical properties of a carbon fibre epoxy composite manufactured by resin transfer moulding using conventional and microwave heating. *Compos. Sci. Technol.* **2008**, *68*, 1854–1861. [CrossRef]
15. Tominaga, Y.; Shimamoto, D.; Hotta, Y. Quantitative evaluation of interfacial adhesion between fiber and resin in carbon fiber/epoxy composite cured by semiconductor microwave device. *Compos. Interfaces* **2016**, *23*, 395–404. [CrossRef]
16. Japanese Standards Association. *Testing Methods for Carbon Fiber Content and Void Content of Carbon Fiber Reinforced Plastics*; Japanese Industrial Standard K7075; Japanese Standards Association: Tokyo, Japan, 1991.
17. Jacquelin, J. A reliable algorithm for the exact median rank function. *IEEE Trans. Electr. Insul.* **1993**, *28*, 168–171. [CrossRef]
18. Pardini, L.C.; Manhani, L.G.B. Influence of the testing gage length on the strength, young's modulus and weibull modulus of carbon fibres and glass fibres. *Mater. Res.* **2002**, *5*, 411–420. [CrossRef]
19. Kelly, A.; Tyson, W.R. Tensile properties of fibre-reinforced metals: Copper/tungsten and copper/molybdenum. *J. Mech. Phys. Solids* **1965**, *13*, 329–350. [CrossRef]
20. Tripathi, D.; Lopattananon, N.; Jones, F. A technological solution to the testing and data reduction of single fibre fragmentation tests. *Compos. Part A Appl. Sci. Manuf.* **1998**, *29*, 1099–1109. [CrossRef]
21. Sui, X.; Shi, J.; Yao, H.; Xu, Z.; Chen, L.; Li, X.; Ma, M.; Kuang, L.; Fu, H.; Deng, H. Interfacial and fatigue-resistant synergetic enhancement of carbon fiber/epoxy hierarchical composites via an electrophoresis deposited carbon nanotube-toughened transition layer. *Compos. Part A Appl. Sci. Manuf.* **2017**, *92*, 134–144. [CrossRef]

22. Yao, T.-T.; Wu, G.-P.; Song, C. Interfacial adhesion properties of carbon fiber/polycarbonate composites by using a single-filament fragmentation test. *Compos. Sci. Technol.* **2017**, *149*, 108–115. [CrossRef]

23. Haider, M.; Hubert, P.; Lessard, L. Cure shrinkage characterization and modeling of a polyester resin containing low profile additives. *Compos. Part A Appl. Sci. Manuf.* **2007**, *38*, 994–1009. [CrossRef]

24. Johnsen, B.; Kinloch, A.; Mohammed, R.; Taylor, A.; Sprenger, S. Toughening mechanisms of nanoparticle-modified epoxy polymers. *Polymer* **2007**, *48*, 530–541. [CrossRef]

25. Kinloch, A.J. *Adhesion and Adhesives: Science and Technology*; Springer: Dordrecht, The Netherlands, 2012.

26. Beyerlein, I.J.; Phoenix, S.L. Statistics for the strength and size effects of microcomposites with four carbon fibers in epoxy resin. *Compos. Sci. Technol.* **1996**, *56*, 75–92. [CrossRef]

27. Sager, R.J.; Klein, P.J.; Lagoudas, D.C.; Zhang, Q.; Liu, J.; Dai, L.; Baur, J.W. Effect of carbon nanotubes on the interfacial shear strength of t650 carbon fiber in an epoxy matrix. *Compos. Sci. Technol.* **2009**, *69*, 898–904. [CrossRef]

28. Decareau, R.V.; Peterson, R.A. *Microwave Processing and Engineering*; E. Horwood: Chichester, England, 1986.

29. Mijović, J.; Wijaya, J. Review of cure of polymers and composites by microwave energy. *Polym. Compos.* **1990**, *11*, 184–191. [CrossRef]

30. Agrawal, R.K.; Drzal, L.T. Effects of microwave processing on fiber-matrix adhesion in composites. *J. Adhes.* **1989**, *29*, 63–79. [CrossRef]

31. Lewis, D.; Summers, J.; Ward, T.; McGrath, J. Accelerated imidization reactions using microwave radiation. *J. Polym. Sci. Part A Polym. Chem.* **1992**, *30*, 1647–1653. [CrossRef]

32. Dai, Z.; Shi, F.; Zhang, B.; Li, M.; Zhang, Z. Effect of sizing on carbon fiber surface properties and fibers/epoxy interfacial adhesion. *Appl. Surf. Sci.* **2011**, *257*, 6980–6985. [CrossRef]

33. Xu, X.; Wang, X.; Cai, Q.; Wang, X.; Wei, R.; Du, S. Improvement of the compressive strength of carbon fiber/epoxy composites via microwave curing. *J. Mater. Sci. Technol.* **2016**, *32*, 226–232. [CrossRef]

34. Zhou, J.; Li, Y.; Li, N.; Hao, X.; Liu, C. Interfacial shear strength of microwave processed carbon fiber/epoxy composites characterized by an improved fiber-bundle pull-out test. *Compos. Sci. Technol.* **2016**, *133*, 173–183. [CrossRef]

35. Shimamoto, D.; Imai, Y.; Hotta, Y. Kinetic study of resin-curing on carbon fiber/epoxy resin composites by microwave irradiation. *Open J. Compos. Mater.* **2014**, *4*, 85–96. [CrossRef]

Article

A New Way of Toughening of Thermoset by Dual-Cured Thermoplastic/Thermosetting Blend

Shankar P. Khatiwada [1,2], Uwe Gohs [2,*], Ralf Lach [3], Gert Heinrich [2,4] and Rameshwar Adhikari [1]

[1]   Research Center for Applied Science and Technology, Tribhuvan University, Kiritipur, Kathmandu 44613, Nepal; skradius33@gmail.com (S.P.K.); nepalpolymer@yahoo.com (R.A.)
[2]   Leibniz-Institut für Polymerforschung Dresden e.V., Hohe Straße 6, 01069 Dresden, Germany; gheinrich@ipfdd.de
[3]   Polymer Service GmbH Merseburg, Eberhard-Leibnitz-Straße 2, 06217 Merseburg, Germany; ralf.lach@psm-merseburg.de
[4]   Institut für Textilmaschinen und Textile Hochleistungswerkstofftechnik, Technische Universität Dresden, 01069 Dresden, Germany
*   Correspondence: gohs-dresden@t-online.de; Tel.: +49-351-4658-239

Received: 21 December 2018; Accepted: 5 February 2019; Published: 12 February 2019

Abstract: The work aims at establishing the optimum conditions for dual thermal and electron beam curing of thermosetting systems modified by styrene/butadiene (SB)-based triblock copolymers in order to develop transparent and toughened materials. The work also investigates the effects of curing procedures on the ultimate phase morphology and mechanical properties of these thermoset–SB copolymer blends. It was found that at least 46 mol% of the epoxidation degree of the SB copolymer was needed to enable the miscibility of the modified block copolymer into the epoxy resin. Hence, an electron beam curing dose of ~50 kGy was needed to ensure the formation of micro- and nanostructured transparent blends. The micro- and nanophase-separated thermosets obtained were analyzed by optical as well as scanning and transmission electron microscopy. The mechanical properties of the blends were enhanced as shown by their impact strengths, indentation, hardness, and fracture toughness analyses, whereby the toughness values were found to mainly depend on the dose. Thus, we have developed a new route for designing dual-cured toughened micro- and nanostructured transparent epoxy thermosets with enhanced fracture toughness.

Keywords: block copolymers; dual curing; electron beam; epoxy resins; toughness

---

1. Introduction

1.1. General Remarks

Thermosetting materials are used as high-performance materials such as adhesives, composites, and coatings in the aerospace and electronics industries [1–4]. However, the high performance of the material is limited by its brittle behavior because of its poor resistance to crack initiation and low fracture energy value [5]. Ineffective compatibilization of thermosetting mixtures of polymers with high interfacial tension often presents poor mechanical properties for the blends. So, for high-performance applications, the toughness of the epoxy thermosets should be increased by improving their resistance to crack propagation without significant drops in other important inherent properties such as modulus, glass transition temperature (T_g), and transparency. To overcome this limitation, many attempts have been successfully completed for thermally cured systems by incorporating the second phase into the matrix of epoxy resins through physical blending or chemical reactions [6–10].

Among some established toughening strategies, the molecular modifications induced by the electron beam (EB) irradiation strategy can strongly alter the mechanical, electrical, and thermal properties of the polymers [11–13]. Therefore, the EB curing is considered as an effective and sustainable method for modification of polymers in comparison to thermal curing [14,15]. The use of EB-modified materials for transport applications has considerably increased in the last few decades for their favorable specific strength and resistance to both corrosion and chemicals as well. The benefits of EB curing for the manufacturing of high-performance materials have been widely discussed, and the importance of this process was suggested by different research groups [12,13,16].

EB curing was successfully tested on acrylic derivatives of epoxies, leading to polymerization via a radical mechanism. However, the final product did not address the required glass transition temperature (T_g), elastic modulus, and fracture toughness for aerospace and advanced automotive applications [15,17–19]. Nevertheless, the enhanced thermal and mechanical properties of the EB-cured materials similar to thermal curing were obtained by the cationic polymerization of epoxy resins using suitable onium salts as initiators [20]. Several parameters, such as composition and structure of the epoxy resin, dose, dose rate, gas atmosphere, and curing temperature, can greatly influence the final properties of the cured materials [21,22]. Epoxy thermoset is prepared by either thermal curing or EB curing. Both curing methods provide high glass transition temperatures and high elastic moduli. EB curing uses accelerated electrons in order to polymerize and crosslink the resins.

1.2. The EB Irradiation Process

EB irradiation is a sustainable technique and has been introduced in the field of material science nearly fifty years ago. The earliest use of this technology was for the sterilization of disposable medical products, conservation of food products, and crosslinking of plastic materials. The curing processes of resin-based materials were developed a bit later [23]. In recent years, the application of this technology has grown more and more. It is widely used to produce heat-shrinkable plastic films, as insulator for electrical wires, and cables in order to increase heat resistance and to enhance the resistance against abrasion and solvents. Besides these applications, EB irradiation can induce the reduction of the molecular mass by scission of the polymers [17]. This leads to the formation of highly applicable engineering polymeric materials which are extensively used in the automotive and aeronautic/aerospace industries [15,18,19,24].

The EB irradiation can be used to produce micro- and nanostructure systems [25]. Accelerated electrons gradually lose their energy through a huge number of small energy transfers. Basically, the effect of EB irradiation depends on several parameters. These are the absorbed dose (D); the energy absorbed per unit of mass, measured in Gray (Gy)), dose rate (energy absorbed per unit of mass and unit time, measured in Gy/s) [26], gas atmosphere, and the constitution of the polymer to be irradiated. The interaction of EB with the polymer causes the formation of excited states or ions and secondary electrons (Scheme 1). These reactive species are converted into free radicals, leading to chain scission, molecular degradation, chain branching, and cross-linking [17]. The ratio of crosslinking to scission reaction depends on the constitution of polymers, as well as the EB treatment conditions. The outlines of the reaction mechanisms are shown in Scheme 1 [27].

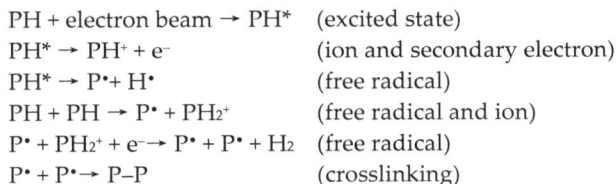

$$PH + electron\ beam \rightarrow PH^* \qquad (excited\ state)$$
$$PH^* \rightarrow PH^+ + e^- \qquad (ion\ and\ secondary\ electron)$$
$$PH^* \rightarrow P^{\bullet} + H^{\bullet} \qquad (free\ radical)$$
$$PH + PH \rightarrow P^{\bullet} + PH_2^+ \qquad (free\ radical\ and\ ion)$$
$$P^{\bullet} + PH_2^+ + e^- \rightarrow P^{\bullet} + P^{\bullet} + H_2 \qquad (free\ radical)$$
$$P^{\bullet} + P^{\bullet} \rightarrow P–P \qquad (crosslinking)$$

Scheme 1. The outline of some reactions illustrating typical physico-chemical transformations on polymers (P—polymer chain).

The irradiation process allows for the curing reactions; in some cases, post-thermal treatments have been subjected in order to complete the curing reaction. It is important to consider that the post-thermal curing is performed on already polymerized materials. The combination of a dual curing process, i.e., radiation curing at moderate temperature, followed by a thermal treatment at high temperature, allows for an improvement of the thermal performance, and significant increase of the elastic modulus. However, the low-fracture toughness originates from the highly crosslinked structure restricting the molecular behavior and can give rise to very brittle materials. Therefore, these materials need to be toughened to enable more applications. The toughness of EB-cured epoxy thermoset can be increased by the use of suitable toughening agents [6–10,13]. A post-irradiation thermal curing allows us to uniform the structure and to obtain a sufficiently high value of T_g, indicating an increase of the crosslinking degree. The presence of the thermoplastic elastomer as toughening agent induces a marked effect on toughness.

EB-cured epoxy resin (80–90 kGy) containing 10 wt.% polyether sulphone (PES) has a value of the elastic modulus of about 3.6 GPa. The critical intensity factors K_{IC}, i.e., the fracture toughness for both diglycidyl ether of bisphenol F (DGEBF) neat resin and DGEBF/PES blends, are reported [28,29]. In addition, the K_{IC} values and the related references of epoxy resins are summarized in the comprehensive data compilation given by Lach and Grellmann [30]. When passing from neat epoxy resin systems to blends, increased K_{IC} values were observed [31].

Recent trends in the global situation towards environmentally conscious manufacturing practices have led us to look for alternative thermal curing technologies. The EB curing is considered as an effective and appropriate method for the modification of polymers in comparison to thermal curing because of its fast curing time, ambient curing temperature, homogeneity, and thermal initiator free curing [14,15]. This modern approach is generally referred to as the green method.

1.3. Motivation

In this paper we are looking for a new strategy to make a toughened and transparent epoxy thermoset for advanced applications. If the epoxy resin was modified by a polymer containing a high amount of polybutadiene (PB), the blend is not transparent and toughened. Here a new approach is described to increase toughness and maintain stiffness and transparency by EB irradiation followed by thermal curing. Finally, the toughened materials are prepared by using small amounts of block copolymers (BCPs), so this method is cost effective as well.

Numerous studies have been carried out to investigate the effects of block copolymer on the mechanical, thermal, and electrical properties of modified epoxy systems [18,19,24,32–35], but the details on the effects of the dose on the ultimate phase morphology and mechanical properties of the dual-cured epoxy thermosetting system modified by block copolymers were not reported. Hence, the aim of this research was to achieve the preparation of toughened dual-cured micro- and nanostructured transparent epoxy thermosets. The ultimate phase morphology and thermo-mechanical properties were analyzed. This research paper also summarizes the important aspects of EB-cured epoxies and their processes.

2. Experimental Details

2.1. Materials and Specimen Preparation

Diglycidyl ether of bisphenol A, provided by Sigma Aldrich, was used as the matrix (Figure 1a), and 4,4-diaminodiphenylmethane (DDM), purchased from Sigma Aldrich, was used as the curing agent (Figure 1c). The commercial styrene/butadiene-based linear triblock copolymer named as Kraton D1101 (31% by volume of styrene) was used as a modifier [34] (Figure 1b). The lab grade chemicals were used without any further purification.

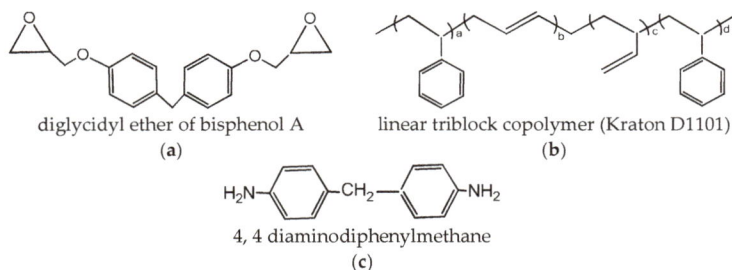

Figure 1. The chemicals used and their structures. (**a**) Diglycidyl ether of bisphenol A (DGEBA); (**b**) linear triblock copolymer (Kraton D1101); (**c**) 4,4 diaminodiphenylmethane (DDM).

2.1.1. Epoxidation of Kraton D1101 by *m*-CPBA

The epoxidation of the styrene/butadiene-based triblock copolymer was accomplished by using m-chloroperoxybenzoic acid (*m*-CPBA) [35]. In a typical procedure, 100 g of Kraton D1101 and 500 mL of dichloromethane were poured into a three-necked round bottom flask and stirred until the polymer was completely dissolved. To achieve a degree of epoxidation (DOE) of around 50 mol%, stoichiometric amounts of *m*-CPBA were charged to the polymer solution. The mixture was then vigorously agitated for two hours under nitrogen gas at 0 °C. After the reaction was completed, the polymer solution was filtered. The excess of *m*-CPBA was removed by extraction with a saturated aqueous sodium bicarbonate (NaHCO$_3$) solution and the mixture was further dried by the sodium sulphate (Na$_2$SO$_4$) solution. The final polymer solution was recovered using a separating funnel and a vacuum suction pump. The solvent was completely evaporated to recover the solid residue which was dried under a vacuum at room temperature. The chemical reaction involved herein is shown in Figure 2. The epoxidized Kraton D1101 is denoted as eKraton.

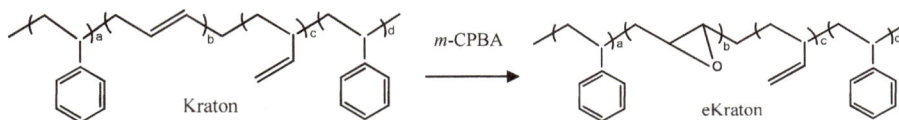

Figure 2. Simplified chemical reactions involved during the epoxidation of the star block copolymer by *m*-CPBA.

The oxygen-oxygen bond in *m*-CPBA is quite weak (about 33 kcal/mol) which leads to the highly reactive nature. One of the prime impacts of this convention is that the stereochemistry is always retained. That means, a cis-alkene always gives the cis-epoxide, and a trans-alkene gives a trans-epoxide product. This is a prime example of a stereoselective reaction in organic chemistry. During the reaction, a very reactive transition state is formed, where the bond between the oxygen and the alkene is being formed at the same time that the O–O bond is breaking, and the proton is being transferred from the OH to the carbonyl oxygen. Those little-dotted lines represent partial bonds. The mechanism of *m*-CPBA/BCPs system is shown in Scheme 2.

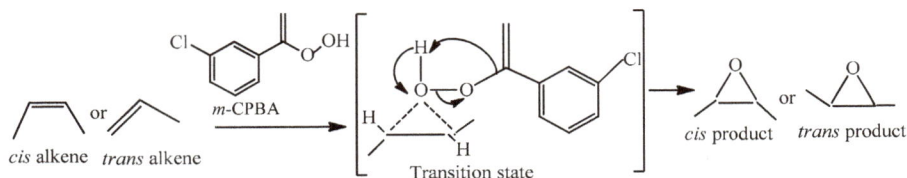

Scheme 2. The mechanism of the *m*-CPBA/BCPs system.

2.1.2. Electron Beam Treatment of EP/eKraton Blends

eKraton was dissolved in excess of dichloromethane and mixed with epoxy resin (EP) in the ratio of 90/10 wt.%. The logic behind taking only 10 wt.% of eKraton was to prevent the reaction mixture from gelation before curing. The mixture was then kept in a vacuum oven to allow the solvent to evaporate. Afterwards, an EB treatment was performed at a temperature of 215 °C under a nitrogen atmosphere with doses of 20 kGy and 50 kGy using an ELV-2 electron accelerator from the Budker Institute of Nuclear Physics (Novosibirsk, Russia) using the procedure we previously reported [36] (the doses were designated as "-20" and "-50", respectively). Then the mixture of EP/eKraton was placed on an electrically-heated plate in an irradiation vessel which was mounted on the conveyor system of the electron accelerator. All treatments were performed with 1.5 MeV at an electron current of 4.0 mA.

2.1.3. Blending of the Epoxidized Mixture

After that, the pure thermoset resin was prepared by the reaction of a stoichiometric amount of EP with DDM (i.e., epoxy resin/DDM in the ratio of 70/30 wt.%). The mixture was pre-cured at 150 °C for 2 h followed by post-curing at 180 °C for another 2 h. The curing of the epoxy system containing 10 wt.% eKraton with doses of 20 kGy and 50 kGy was carried out as described in Reference [36]. In the case of a typical EB and thermal cured system, first the EB-cured EP/eKraton blends were mixed in a stoichiometric amount of DDM and magnetically stirred over a hot plate at 100 °C until the complete dissolution of DDM leading to the formation of a homogenous mixture. The same thermal curing profile has followed, as in the case of pure thermoset resins. Then, each thermally-cured blend was cooled gradually down to room temperature. Finally, the prepared pure epoxy thermoset and the blend were cut into proper sizes according to the requirements as demanded by different characterization techniques.

2.2. Characterization Techniques

2.2.1. Molecular Characterization

The ^{1}H-NMR spectra (proton nuclear magnetic resonance) were recorded using a Bruker Advance III 500 spectrometer (Bruker, Billerica, MA, USA) operating at 500.13 MHz for ^{1}H. The samples were measured at 30 °C in deuterated chloroform (CDCl$_3$) which was also used as a reference (δ (^{1}H) = 7.26 ppm).

2.2.2. Morphological Characterization

Optical microscopy (OM): EP/eKraton blends were inspected using a Nikon Optiphot-2 optical microscope (Nikon, Minato, Japan) in transmission mode before and after curing. Small droplets of blends were sandwiched between two clean and dry glass coverslips, slightly pressed against each other, and the morphology was observed at room temperature.

Scanning electron microscopy (SEM): SEM was used for the study of the morphological features and deformation behaviors of the materials. The fracture surface of fractured materials obtained from impact test was sputter coated with an approximately 20–50 nm thin film of gold film. The thin film has two functions: Avoiding surface charging and irradiation damage. The experiments were carried out by using scanning electron microscopy (JSM 6300, JEOL, Akishima, Japan) employing secondary electrons (SE) mode.

Transmission electron microscopy (TEM): TEM was used to investigate the phase morphology of the specimens using a transmission electron microscope LIBRA120 (Carl Zeiss AG, Oberkochen, Germany) with an acceleration voltage of 120 kV. Cured bulk thermoset samples were prepared by using a Leica UCT ultramicrotome equipped with a diamond knife with a cutting edge of 45°. Ultra-thin cross-sections of ~60 nm thickness were cut at −80 °C from the sample with an ultra-microtome EM UC6/EM FC6 (Leica Microsystems, Austria) using a diamond knife and were subsequently transferred to carbon-coated grids. The soft polybutadiene (PB)-rich phase was stained with osmium tetroxide to enhance contrast before the TEM imaging.

2.2.3. Mechanical Testing

Impact strength: The impact strength of the blends was measured using a Charpy impact tester PSW 4 J (Polymer Service GmbH Merseburg, Merseburg, Germany) according to ISO 179-1. The sample dimensions were 100 mm × 10 mm × 3 mm. The notched specimen would be broken with one blow of the hammer. The Charpy impact strength (in kJ/m^2) was calculated as the ratio of the fracture energy (in J) and the cross-sectional area (in mm^2). Five successive measurements of each blend were performed, and their mean average result was taken into account.

Fracture toughness: The fracture toughness testing of the blends was conducted using a Zwick universal testing machine Z2.5 (ZwickRoell GmbH & Co. KG, Ulm, Germany) according to ISO 13586 in three-point bending mode. Single edge-notched bend (SENB) samples having a dimension (L × W × B) of 55 mm × 12 mm × 5.85 mm were prepared. Then, a notch with a depth of a = 5.6 mm was initiated by tapping a fresh razor blade. The critical plane strain energy release rate (G_{IC}) was determined from the critical plane strain stress intensity factor (K_{IC}) obtained. Care was taken to avoid forming a long crack or breaking the sample. Specimens with long cracks, i.e., a/W exceeding 0.55, were discarded.

Microindentation: The microhardness measurements were carried out at 23 °C by using FischerscopeH100C recording microhardness tester (Helmut Fischer GmbH, Sindelfingen, Germany) equipped with a pyramidal Vickers diamond indenter which was indented into the sample. This technique comprises the continuous measurement of the load applied by an indenter as a function of its indentation depth with the same speed (50 μm/s) for both loading and unloading. The hardness has been measured in the form of the Martens hardness HM (in N/mm^2) from the maximum load (in N) and the modulus of elasticity (the indentation modulus E_{IT}) has been calculated according to the procedure well-described in ISO 14577. At least five load (F)–indentation depth (h) measurements per material with the maximum load of F_{max} = 1 N were done at different positions on the material plate of approximate area 1 × 1 cm^2 and thickness around 3 mm and the average value was reported.

2.2.4. Dynamic Mechanical Analysis

Dynamic mechanical analysis (DMA): DMA measurements were carried out on the specimens with dimensions of 30 mm × 10 mm × 3 mm using a DMA Q800 V21.1 rheometer (TA Instruments, New Castle, GB) in torsion mode. The dynamic mechanical properties of samples were measured over a broad temperature range (from −70 to 200 °C) at a frequency of 1 Hz. The temperature dependency of storage modulus (G'), loss modulus (G'') and mechanical loss factor (tan δ = G''/G') were measured.

3. Results and discussion

3.1. Quantitative Validation of the Chemical Modification

Figure 3 depicts a region of the ^{1}H-NMR spectra of Kraton (bottom) and partially epoxidized Kraton (top). The signals in the 7.2–6.2 ppm region result from the aromatic ring of the styrene units. They remain unchanged in the epoxidation reaction and their intensity can be used as internal intensity reference to compare the spectrum of a non-epoxidized and epoxidized sample. The signals of the olefinic protons of the butadiene segments appear in the 5.8–4.8 ppm region. The copolymerization with butadiene can result in *cis*-and *trans*-1,4-butadiene units, respectively, and in vinylic 1,2-butadiene units. The ^{1}H chemical shifts of the corresponding protons are summarized in Table 1.

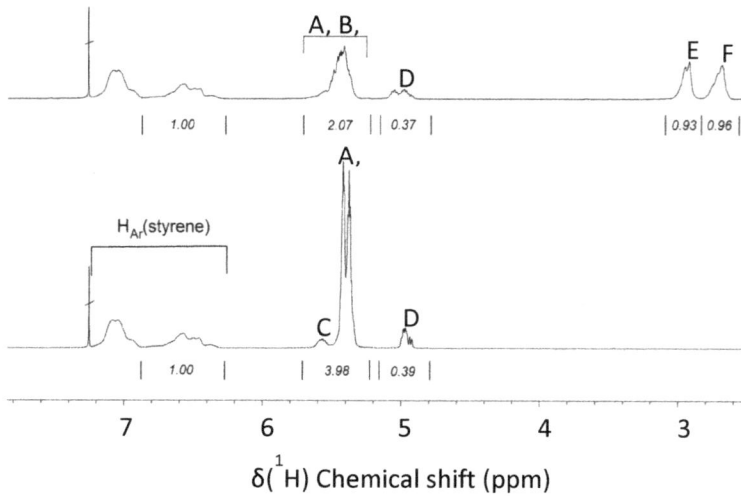

Figure 3. Spectra of eKraton (**top**) compared with Kraton D1101 (**bottom**); solvent: CDCl₃.

Table 1. Assignment of chemical shift values (δ) in the ¹H-NMR spectrums of the virgin Kraton and eKraton [37].

δ (ppm)	Protons	Remarks
5.5	–CH₂–CH=CH–CH₂–	*cis* hydrogen of 1,4-butadiene
5.4	–CH₂–CH=CH–CH₂–	*trans* hydrogen of 1,4-butadiene
5.0	–CH=CH₂	vinyl hydrogen of 1,2-butadiene
5.5	–CH=CH₂	vinyl hydrogen of 1,2-butadiene
2.7	⌃O⌃ –CH—CH–	*trans*-epoxy
2.9	⌃O⌃ –CH—CH–	*cis*-epoxy

Whereas the signals of *cis*- (A) and *trans*-1,4-butadiene units (B) almost overlap at about 5.5 and 5.4 ppm, the –CH= (C) and =CH₂ signals (D) of the vinyl-1,2-butadiene units appear well separated at about 5.6 and 5.0 ppm, respectively. Epoxidation results in decreasing intensity of the *cis*- and *trans*-1,4-butadiene signals, but the signal for the 1,2-butadiene unit of Kraton remains almost unchanged. The epoxidation of vinyl–1,2-butadiene was observed in the ¹H-NMR spectrum only to a low extent, which could be due to the lower reactivity of vinyl bonds in comparison to *cis* and *trans* double bonds, which is also reflected in the Fourier-transform infrared (FTIR) data.

The epoxide groups in the eKraton result in new signals at about 2.9 and 2.7 ppm, representing the *cis*- (E) and *trans*-epoxy groups (F), respectively. The overall signal patterns in the olefinic and epoxide signal regions strongly depend on the degree of epoxidation (DOE). Adjusting the same region of the aromatic styrene signals to the same intensity, the integral values of the rubbery part before and after epoxidation can be compared. The decrease in the intensity of the peaks corresponding to olefinic protons is related to the DOE.

The *cis*- and *trans*-1,4-butadiene units contain two olefinic protons, but the 1,2-butadiene units contain three olefinic protons. This has to be considered in the calculation. Therefore, the overall integral of the olefinic protons is corrected by half the intensity of the signal group at 5.0 ppm. The following Equation (1) is used to determine the DOE.

$$\text{DOE} = 1 - \frac{\text{Corrected integral area of olefinic protons after epoxidatio n}}{\text{Corrected integral area of olefinic protons before epoxidatio n}} \times 100 \, \text{mol\%} \qquad (1)$$

The calculation is based on the integral values of the spectra (see Figure 3). The corrected integral area of the olefinic protons before epoxidation amounts to 4.175 (3.98 + 0.39 − 0.195). In contrast, the corrected integral area of olefinic protons after epoxidation amounts to 2.255 (2.07 + 0.37 − 0.185). Consequently, the DOE was 46 mol%.

3.2. Preliminary Morphological Analysis

The photographs of the epoxy thermosets (EP) and their blends fabricated with 10 wt.-% of eKraton and treated with different doses are presented in Figure 4.

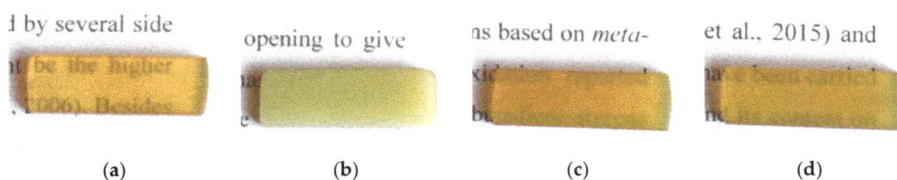

(**a**)	(**b**)	(**c**)	(**d**)

Figure 4. Photographs showing the different levels of optical transparency of EP/eKraton blends containing 10 wt.% of eKraton of different irradiation doses in the visible light range (400–700 nm). (**a**) Pure epoxy (EP); (**b**) eKraton; (**c**) eKraton-20; (**d**) eKraton-50 (the width of the sample is 10 mm in each case).

All cured blends were not transparent as targeted. Obviously, the pure EP thermoset was transparent (Figure 4a) because the epoxy system consists of a single phase. After the addition of 10 wt.% eKraton modifiers into the epoxy system (EP/eKraton), the blend turned opaque (Figure 4b), possibly due to the macrophase separation of eKraton in the EP matrix. The use of 10 wt.% eKraton was not enough to make the PB block completely compatible with the epoxy matrix. After an EB treatment with a dose of 20 kGy (EP/eKraton-20), the EB-modified blend became more translucent as compared with non-irradiated EP/eKraton, as shown in Figure 4c.

This implies the formation of micro/nanosized domains of eKraton-20 in the blend. As the EB dose further increases to 50 kGy, the blend was almost transparent, as shown in Figure 4d, suggesting the formation of nanostructures in eKraton-50 along with some microstructures. In the case of transparent micro- and nanostructured blends, the material allows the light to pass because the size of the majority of the domain is less than 700 nm. Similar results were also observed in the epoxy/styrene–butadiene–styrene block copolymer (SBS) system, where epoxidized SBS induced the nanostructure in epoxy resin and the transparency of EP/SBS largely depended on the degree of epoxidation [33,38]. For more information about the morphology, the blends were further analyzed using optical microscopy (OM) and transmission electron microscopy (TEM).

3.3. Morphological Behavior

Optical micrographs of the blends of EP/eKraton (90/10 wt.%) before and after curing are displayed in Figure 5. It was observed that the partially epoxidized (46 mol%) ePB subchains of eKratons (dispersed phase) were initially miscible with the epoxy matrix (continuous phase), whereas the immiscible PS subchains were arranged in the micrometer range. The epoxy resin (EP) acts as the selective solvent for the eKratons and it manages the driving force for SA mechanisms. The size of the domains was found to be around 3 μm.

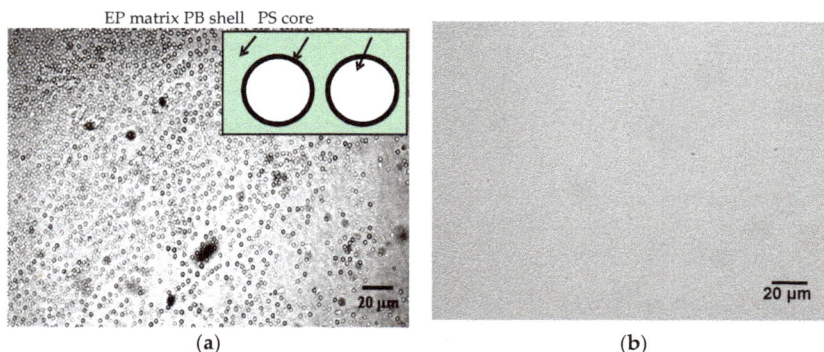

Figure 5. Optical photographs showing the morphology of EP/eKraton (90/10 wt.%): (**a**) Before and (**b**) after curing.

The epoxidized butadiene segments were miscible with the epoxy resin, whereas polystyrene parts as well as non-epoxidized parts were immiscible. The non-epoxidized butadiene (PB) parts formed the shell, and polystyrene (PS) parts may formed the core of the domain, as shown in the schematic enlargement presented in the inset of Figure 5a. However, the microstructure of eKraton/EP after curing was not observed (Figure 5b), because the majority of the microdomains of polystyrene were converted into nanodomains, and the nanostructure cannot be observed by optical imaging.

For detail analysis of the size and distribution of the domain in the nanoscale range, TEM analysis were carried out. The EP/Kraton blend morphology with 10 wt.% of unmodified block copolymer is presented in Figure 6a, whereby the morphology of the EP/eKraton blend treated by a different dose is shown in Figure 6b,c. At first glance, heterogeneous morphology was seen in all cases. It is noted that the EP/eKraton blend, which was not cured by EB irradiation, consisting of a bright spherical domain in the 0.5–1 µm range, dispersed in the continuous epoxy matrix, as indicated in light gray. Owing to the electron density difference of different groups and the diversity in the preference of RuO_4 staining, the continuous areas can be ascribed to the cross-linked epoxy matrix, and the spherical cores as indicated in dark represent the polystyrene parts along with block of non-modified rubbery polybutadiene. The lamellar morphology was also observed inside the spherical core, where light areas correspond to the styrene block, whereas the thin dark shells surrounding the spherical cores can be assigned to the non-epoxidized PB segment.

The average sizes of the spherical microdomains in the EP/eKraton blends were estimated to be 0.5–1 µm using the measuring tool of the software Image J (Laboratory for Optical and Computational Instrumentation (LOCI)). The size of the majority of eKraton domains was larger than the wavelength of light (700 nm). This is the reason why the blend appears opaque, which was in good agreement with the photograph as described in Figure 4b. As the dose increases, the spherical morphology consisting of a bright core and a dark shell remains unchanged; however, the size of the domains were decreased gradually, the number of spherical domains was increased, and the distance between them decreased.

All these results indicate the higher miscibility of eKraton with epoxy resin as the dose of EB treatment increases. Obviously, the EB treatment induced a crosslinking of the ePB subchains with epoxy resin before thermal curing. We conclude that only EB treatment was responsible for the changes in size, positions, and distribution of the domains, since other controlling conditions like post-thermal temperature, hardener, and curing conditions remained constant. In Figure 6b, the sizes of the domains were found to be in the nm−µm range. However, the majority of the domains were in the nm range, and their sizes were comparable with the wavelength of visible light, and hence this blend was translucent in nature, indicating that not all domains were on the nanoscale. These results are in good agreement with the photographs described in Figure 4c. Furthermore, the blends treated by a dose of 50 kGy contain the dispersed domains with sizes in a 50–100 nm scale, as shown in Figure 6c. Since the domain sizes were comparable with the wavelength of visible light, mainly the EP/eKraton-50 blends were transparent. This result is also supported by the image of the micrograph as shown in Figure 4d.

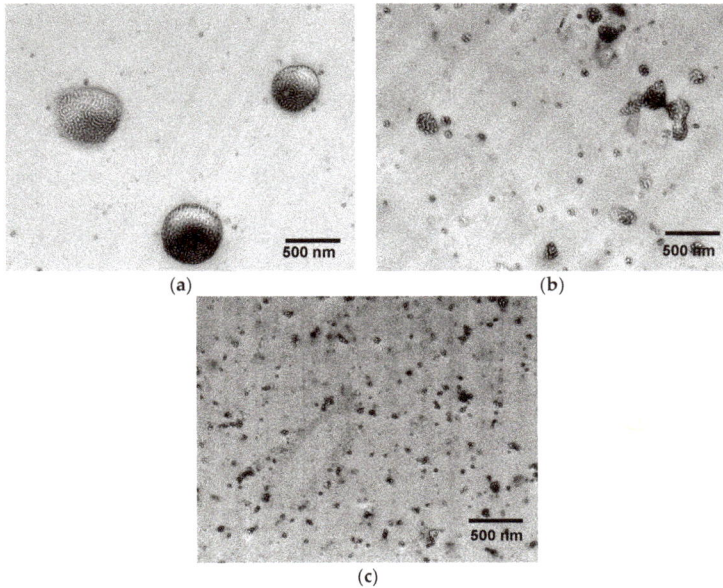

Figure 6. TEM images of EP/eKraton blends (90/10 wt.%) with different irradiation doses. (**a**) EP/eKraton; (**b**) EP/eKraton-20; and (**c**) EP/eKraton-50 at 0, 20 and 50 kGy, respectively.

3.4. Dynamic Mechanical Behavior

Figure 7 shows the temperature dependencies of the dynamic mechanical properties (storage modulus (G′) and damping coefficient (tan δ)) of the pure epoxy thermoset and EP/eKraton blends with different doses. In general, the addition of modified block copolymers (eKraton) diminishes the storage modulus of epoxy resin due to the presence of a soft rubbery phase at high temperatures (T). Here, the storage modulus for EP/eKraton treated by EB increases for T < 100 °C and decreases for T > 100 °C. The values obtained from DMA are listed in Table 2. The unmodified epoxy has a main glass (α) transition temperature T_g = 173 °C and a secondary (β transition) temperature of −40 °C [39,40].

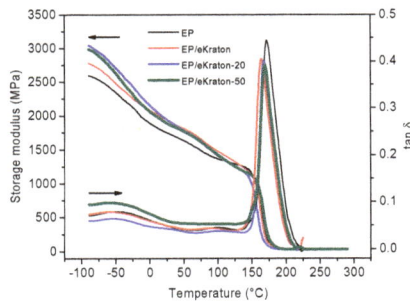

Figure 7. Storage modulus and mechanical loss factor as a function of temperature.

Table 2. Storage modulus and transition temperature values obtained from DMA curves.

Notation	Storage Modulus (G′ in MPa) at 50 °C	β Transition Temperature in °C	α Transition Temperature (T_g in °C)
EP	2467	−40	173
EP + eKraton	2349	−46	164
EP + eKraton-20	2641	−47	168
EP + eKraton-50	2720	−48	170

All blends fabricated with 10 wt.% of eKraton with different doses indicated the two transitions (α and β) with slight changes in the storage modulus (G′) and the peaks of the mechanical loss factor (tan δ). Apart from glass transition temperature, the tan δ peak at T_g also provides some information about the structure of the crosslinked thermoset provided by its shape. Specifically, the height of the tan δ peak at T_g indicates the extent of the crosslinking. A lower height means a higher crosslink density [41]. Moreover, a broader peak suggests better fracture toughness and a better ability to prevent crack propagation and brittle fracture [42].

There is no notable difference on the shape of the tan δ peak at T_g among all EP/eKraton samples treated by different doses. Interestingly, tan δ, which is the ratio of the loss modulus to the storage modulus shows, a significant reduction in T_g (9 K) from EP to EP/eKraton, and an increase (4–6 K) due to irradiation (Table 2). The storage modulus at 50 °C and the β transition temperature have the same trend as found for T_g (Table 2). Similar findings have been also reported by Vignoud [43], as irradiation results in a slight decrease of the T_g and storage modulus in the rubbery region, ascribed to crosslinks.

3.5. Mechanical Behavior

The mechanical properties of the blends are analyzed by subjecting the blends to undergo high-speed impact tests and low-speed fracture toughness and micro-indentation measurements. In Figure 8, the impact strengths of EP/eKraton blends modified with varied doses compared to that of neat epoxy thermoset are presented. In the case of the blends with eKraton, the impact strength increases with dose amount of eKraton. The impact strength of modified epoxy thermosets increases by 25 % for both EP/eKraton-20 and EP/eKraton-50. However, there is no change in the Charpy impact strength for the unmodified epoxy thermoset EP/eKraton within the experimental uncertainty. The good compatibility and strong interaction between the epoxy matrix and epoxidized polybutadiene (ePB) subchains of eKraton after EB irradiation contributed to the improvement of the impact strengths of the blends. However, the value was kept constant in the case of EP/eKraton. This might be because of heterogeneous distribution of the microsized domain, which may facilitate crack initiation and propagation during sudden impact deformation. The results are in good agreements with the morphology (Figure 4) and TEM analyses (Figure 6).

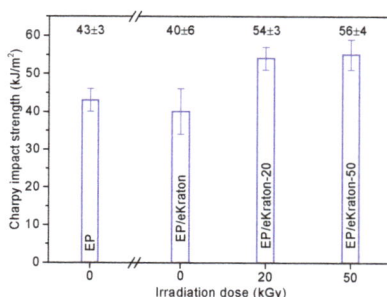

Figure 8. The Charpy impact strengths of EP/eKraton blends modified with different irradiation doses compared with neat epoxy thermoset.

The Charpy impact tests give an idea about the macroscopic mechanical behavior of the blends, while the micromechanical behavior can be evaluated by fracture toughness measurements. Five successive measurements of each blend were performed, and their mean average result was taken into account. Thus, the latter properties of eKraton-modified epoxy thermosets were analyzed by low-speed fracture toughness in order to determine the critical strain energy release rate (G_{IC}) and stress intensity factor (K_{IC}). The fracture toughness values (G_{IC} and K_{IC}) of the blends modified by different doses containing same amount of eKraton (10 wt.%) were compared with pure epoxy thermoset as shown in Table 3.

Table 3. Critical strain energy release rate (G_{IC}), critical stress intensity factor (K_{IC}), and modulus of elasticity (E) of the neat epoxy (EP) thermoset compared with EP/eKraton blends prepared with different irradiation doses (mean values and standard deviation).

Notation	Dose (kGy)	G_{IC} (kJ m^{-2})	K_{IC} (MPa m$^{1/2}$)	E (GPa)
EP	0	0.68 ± 0.02 (G_{IC}^0)	1.46 ± 0.10 (K_{IC}^0)	3.10 (E_0)
EP/eKraton	0	0.67 ± 0.09	1.39 ± 0.10	2.90
EP/eKraton-20	20	1.00 ± 0.06	1.62 ± 0.10	2.65
EP/eKraton-50	50	1.47 ± 0.17	2.11 ± 0.10	2.64

The G_{IC} value of the unmodified epoxy resin was 0.68 kJ m^{-2}. Compared to neat epoxy, the same G_{IC} value was observed for the EP/eKraton within the experimental uncertainty range. This might be due to the heterogeneous distribution of microsized domains, which may facilitate crack initiation and propagation during fracture deformation. However, the fracture toughness values of modified epoxy blends were higher than those of the neat epoxy blends and they increase gradually with increasing dose levels. These results also support the findings obtained from the impact strength analysis (Figure 8).

The G_{IC} values were increased for EP/eKraton-20 and EP/eKraton-50 thermosets by 1.5 and 2.2 times, respectively, as compared to the fracture energy of neat epoxy thermoset. Compared to the neat epoxy thermoset, the nano- and microsized eKratons were capable of toughening the epoxy matrix more effectively due to the strong interaction between the epoxy matrix and ePB sub-chains of eKraton after EB irradiation, as well as the proper dispersion of the micro/nanodomains and improvement of the fracture toughness by resisting the crack initiation and crack propagation. Moreover, the increased toughness was an advantage of the dual curing of the blends via EB irradiation and post-thermal curing in the presence of the small loading of BCPs with comparison to traditional thermal curing.

The dose is the controlling parameter for electron induced reactions as well as the size and distribution of the microdomains in the EP/eKraton blends (see Figure 6). Consequently, the crosslinking behavior of Kraton D1101 was studied at a temperature of 215 °C in a nitrogen atmosphere for different dose values (50, 100, 200 kGy) in Reference [36]. In Reference [36], the dominance of the crosslinking reaction over chain scission was confirmed. Based on those results, the dose values of 20 and 50 kGy were precisely selected. The preparation of blends using a dose > 50 kGy were not successful due to the high values of gel content (>89%) of Kraton D1101. The formations of micro- and nanostructures inside epoxy thermosets were also reported to improve the fracture toughness significantly [33,44–47]. However, the actual mechanism of this toughening is still a subject of debate with respect to whether the modifier or the matrix absorb the most energy during deformation.

Indentation techniques are among the simplest, most reliable, and most efficient methods of testing the surface mechanical properties of thermosetting materials. (Micro)indentation testing has been employed in determining the fracture toughness as well the hardness and elastic modulus of polymeric materials such as micro- and nanostructured thermosets [48,49].

The indentation behavior of the materials can be directly linked to the internal morphologies and deformation behaviors of the materials [49]. In order to determine the effects of the differently-modified block copolymer architectures on the surface hardness of block copolymer/EP, the corresponding values of Martens hardness (HM), maximum indentation depth (h_{max}), and indentation modulus (E_{IT}) are plotted in Figure 9. A closer look at the curves reveals that the h_{max} values of the blends were not changed within the experimental uncertainty range. Compared to the pure epoxy thermoset, the addition of 10 wt.% of eKraton decreased the HM value independently of the dose. This might be attributable to the soft phase. In the case of E_{IT}, the value was decreased by the added eKraton and increased with the dose in order to reach the value of the virgin EP at a dose of 50 kGy. Earlier studies on similar systems show that the mechanisms for epoxies toughened by block copolymers are similar to those of conventional rubber-toughened epoxy, which are crack pinning [50], multiple crazing [51], and debonding [52] mechanisms. Figure 10 shows the fracture surface of epoxy blends with the 10 wt.% block copolymer modified by a dose of 50 kGy (Figure 10b) compared with the brittle surface of the pure epoxy thermoset (Figure 10a). The micrograph reveals the existence of tiny cracks, along the main crack growth direction, and these cracks are pinned by the micro- and nanodomains, as indicated by (A) in Figure 10b.

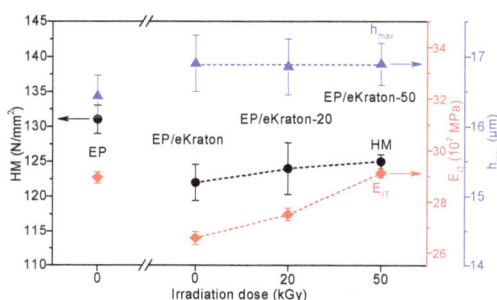

Figure 9. Hardness (HM), indentation modulus (E_{IT}), and maximum indentation depth (h_{max}) of EP/eKraton blends modified with different irradiation doses compared with the neat epoxy thermoset.

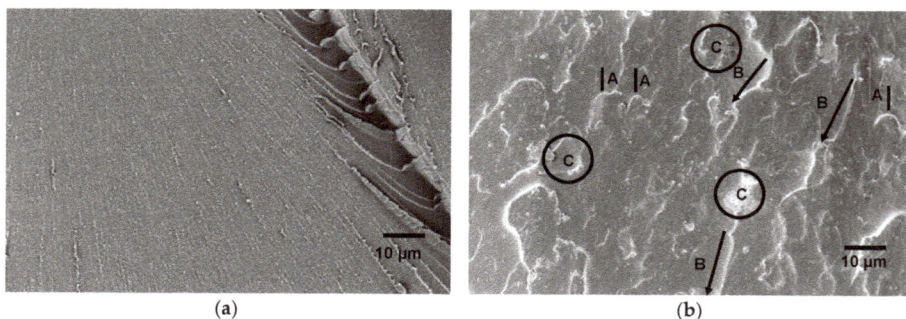

(a) (b)

Figure 10. SEM images showing the fracture surface morphology of EP/eKraton-50 blend (**b**) compared with that of pure epoxy thermoset (**a**); crack pinning (**A**), crack bifurcation (**B**), and debonding (**C**).

In Figure 10b, the fracture surface was similar to that of ductile material due to the plasticization effect of epoxidized PB phases [53]. During this type of deformation, the load was transferred more effectively to the nanostructured domains from the crosslinked epoxy phase [54]. Likewise, bifurcations of microcracks were observed which result in an increased surface area of cracks, as visualized in (B) in Figure 10b. Obviously, this may explain the toughness increase of the material. The energy dissipation mechanisms could also be related to the formation of shear deformation of eKraton domains in the micro- and nanodomains of epoxy thermosets by a debonding process, as indicated by (C) in Figure 10b.

3.6. The Structure—Properties Correlations Scheme

The concept of creating the micro- and nanostructure in the EP/eKraton blends for the construction of toughened thermosets by dual curing methods has been illustrated in Figure 11. When one of the modified blocks of the segments of eKraton is miscible with the epoxy resin (see Figure 11), even in the curing process, and the other segments of non-epoxidized PB segments and PS are immiscible with the epoxy, the ability of self-assembly for Kraton forms the microdomain of eKraton in the epoxy system before curing. However, micro- and nanodomains were formed after dual curing (EB irradiation and subsequent thermal curing) by a reaction-induced phase separation mechanism. Therefore, the phase size of the polymer blends can be controlled in the order of micro- and nanometers. A summary of the structural properties relationship is also shown in Figure 11.

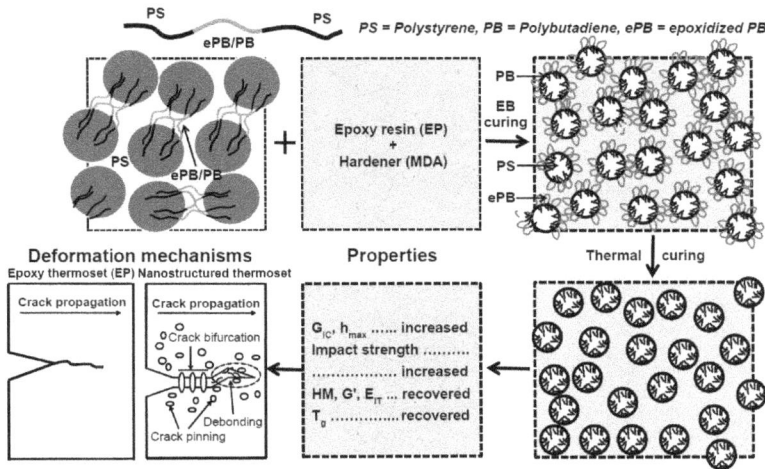

Figure 11. A schematic illustration of the formation of micro/nanodomains in the EP/eKraton blends. The ePB segments of eKraton are miscible with the epoxy system, and the PB and PS segments are immiscible with the DGEBA/MDA system. The micro- and nanostructured EP/eKraton blends are fixed by dual curing, the modulus and stiffness are recovered fractured toughness, impact strength increases, and T_g values decrease. The deformation mechanism follows the crack pinning, crack bifurcation, and debonding processes.

4. Conclusions

The partial epoxidation of a commercial styrene/butadiene-based triblock copolymer Kraton D1101 was achieved and, hence, a dual-cured toughened micro- and nanostructured transparent epoxy thermoset was developed. Some of the important conclusions are given below.

- The quantitative validation of the partial epoxidation of Kraton D1101 was achieved using ^{1}H-NMR spectroscopy.
- The experimental condition for the novel dual curing method was established by optimizing the curing condition and deformation behavior.
- The formation of micro- and nanostructures could be observed, and the effect of dual curing on the ultimate phase morphology was investigated using DMA and TEM analysis.
- The mechanical properties of the modified blends were improved as per the impact strength, fracture toughness, and micro-indentation analyses. The fracture toughness values were found to mainly depend on the dose.

Therefore, the present study has provided an idea for a new toughening strategy for the construction of toughened micro- and nanostructured thermosets. The new materials provide possible uses of the toughened thermosets for advanced industrial applications.

Author Contributions: Conceptualization, R.A., U.G., and G.H.; Investigation: S.P.K.; Methodology, S.P.K. and R.A.; Project Administration, U.G.; Resources, G.H.; Supervision, U.G. and G.H.; Validation, S.P.K., R.A., and U.G.; Visualization, S.P.K. and R.L.; Writing–Original Draft Preparation, S.P.K.; Writing–Review and Editing, U.G., R.A., G.H. and R.L.

Funding: This research was funded by the German Academic Exchange Service (DAAD) within the sandwich PhD scholarship of S.P.K.

Acknowledgments: We are grateful to the German Academic Exchange Service (DAAD) for providing the sandwich PhD scholarship to S.P.K. Finally, the authors would like to thank Sabine Krause and Holger Scheibner from Leibniz-Institut für Polymerforschung Dresden e.V., Germany for rheological studies and thermal analysis, as well as some mechanical measurements. The authors would like to thank Sigma Aldrich and Kraton Polymers International Ltd. for supplying the raw materials.

Conflicts of Interest: The authors declare no conflict of interest.

References

1. Nikafshar, S.; Zabihi, O.; Hamidi, S.; Moradi, Y.; Barzegar, S.; Ahmadi, M.; Naebe, M. A renewable bio-based epoxy resin with improved mechanical performance that can compete with DGEBA. *RSC Adv.* **2017**, *7*, 8694–8701. [CrossRef]
2. Liu, Y.; Zhao, J.; Zhao, L.; Li, W.; Zhang, H.; Yu, X.; Zhang, Z. High performance shape memory epoxy/carbon nanotube nanocomposites. *ACS Appl. Mater. Interfaces* **2015**, *8*, 311–320. [CrossRef] [PubMed]
3. Chen, C.F.; Iwasaki, S.; Kanari, M.; Li, B.; Wang, C.; Lu, D.Q. High performance UV and thermal cure hybrid epoxy adhesive. *IOP Conf. Ser.: Mater. Sci. Eng.* **2017**, *213*, 012032. [CrossRef]
4. Fischer, F.; Beier, U.; Wolff-Fabris, F.; Altstädt, V. Toughened high performance epoxy resin system for aerospace applications. *Sci. Eng. Compos. Mater.* **2011**, *18*, 209–215. [CrossRef]
5. Ellis, B. *Chemistry and Technology of Epoxy Resins*; Chapman & Hall: London, UK, 1994.
6. Unnikrishnan, K.P.; Thachil, E.T. Toughening of epoxy resins. *Des. Monomers Polym.* **2006**, *9*, 129–152. [CrossRef]
7. Blanco, I.; Cicala, G.; Faro, C.L.; Recca, A. Development of a toughened DGEBS/DDS system toward improved thermal and mechanical properties by the addition of a tetra functional epoxy resin and a novel thermoplastic. *J. Appl. Polym. Sci.* **2003**, *89*, 268–273. [CrossRef]
8. Park, S.J.; Jin, F.L. Improvement in fracture behaviors of epoxy resins toughened with sulfonated poly(ether sulfone). *Polym. Degrad. Stab.* **2007**, *92*, 509–514. [CrossRef]
9. Gan, W.; Yu, Y.; Wang, M.; Tao, Q.; Li, S. Morphology evolution during the phase separation of polyetherimide/epoxy blends. *Macromol. Rapid Commun.* **2003**, *24*, 952–956. [CrossRef]
10. Khatiwada, S.P.; Thomas, S.; Saiter, J.M.; Lach, R.; Adhikari, R. Mechanical and thermal properties of triblock copolymer modified epoxy resins. *Bibechana* **2018**, *16*, 196–203. [CrossRef]
11. Jansen, J.U.; Machado, L.D.B. A new resin for photo curable electrical insulating varnishes. *Nucl. Instrum. Methods Phys. Res. B* **2005**, *236*, 546–551. [CrossRef]
12. Alessi, S.; Dispenza, C.; Fuochi, P.G.; Corda, U.; Lavalle, M.; Spadaro, G. E-beam curing of epoxy-based blends in order to produce high-performance composites. *Radiat. Phys. Chem.* **2007**, *76*, 1308–1311. [CrossRef]
13. Alessi, S.; Dispenza, C.; Spadaro, G. Thermal properties of E-beam cured epoxy/thermoplastic matrices for advanced composite materials. *Macromol. Symp.* **2007**, *247*, 238–243. [CrossRef]
14. Naskar, K.; Gohs, U.; Wagenknecht, U.; Heinrich, G. PP–EPDM thermoplastic vulcanisates (TPVs) by electron induced reactive processing. *Express Polym. Lett.* **2009**, *3*, 677–683. [CrossRef]
15. Jhonson, M.A. Electron Beam Processing for Advanced Composites. Available online: https://www.radtech.org/images/pdf_upload/ElectronBeamProcessingforAerospaceComposites.pdf (accessed on 21 December 2018).
16. Lopata, V.J.; Sidwell, D.R. Electron beam processing for composite manufacturing and repair. *Radtech Rep.* **2003**, *17*, 32–42.
17. Woods, R.J.; Pikaev, A.K. *Applied Radiation Chemistry: Radiation Processing*; John Wiley and Sons: New York, NY, USA, 1990.

18. Goodman, D.L.; Palmese, G.R. Curing and bonding of composites using electron beam. In *Handbook of Polymer Blends and Composites*; Kulshresshtha, A.K., Vasale, C., Eds.; Rapra Technology: Shrewsbury, UK, 2002; pp. 459–500.

19. Berejka, A.J.; Montoney, D.; Cleland, M.R.; Loiseau, L. Radiation curing: Coatings and composites. *Nucleonika* **2010**, *55*, 97–106.

20. Bulut, U.; Crivello, J.V. Investigation of the reactivity of epoxide monomers in photo-initiated cationic polymerization. *Macromolecules* **2005**, *38*, 3584–3595. [CrossRef]

21. Nho, Y.C.; Kang, P.H.; Park, J.S. The characteristics of epoxy resin cured by γ-ray and E-beam. *Radiat. Phys. Chem.* **2004**, *71*, 243–246. [CrossRef]

22. Raghavan, J. Evolution of cure, mechanical properties, and residual stress during electron beam curing of a polymer composite. *Compos. A: Appl. Sci. Manuf.* **2009**, *40*, 300–308. [CrossRef]

23. Cleland, M.R.; Parks, L.A.; Cheng, S. Applications for radiation processing of materials. *Nucl. Instrum. Methods Phys. Res. B* **2003**, *208*, 66–73. [CrossRef]

24. Clough, R.L. High-energy radiation and polymers: A review of commercial processes and emerging applications. *Nucl. Instrum. Methods Phys. Res. B* **2001**, *185*, 8–33. [CrossRef]

25. Chmielewski, A.G. Chitosan and radiation chemistry. *Radiat. Phys. Chem.* **2010**, *79*, 272–275. [CrossRef]

26. Adliene, D.; Adlyte, R. Dosimetry principles, dose measurements and radiations protection. In *Applications of Ionizing Radiation in Materials Processing*; Sun, Y.X., Chmielewski, A.G., Eds.; Institute of Nuclear Chemistry and Technology: Warszawa, Poland, 2017; pp. 55–80. ISBN 978-83-933935-8-9.

27. Hafezi, M.; Khorasani, S.N.; Ziaei, F.; Azim, H.R. Comparison of physicomechanical properties of NBR–PVC blend cured by sulfur and electron beam. *J. Elastomers Plast.* **2007**, *39*, 151–163. [CrossRef]

28. ASTM D 695-02a. *Standard Test Methods for Compressive Properties of Rigid Plastics*; ASTM: West Conshohocken, PA, USA, 2002.

29. Alessi, S.; Conduruta, D.; Pitarresi, G.; Dispenza, C.; Spadaro, G. Hydrothermal ageing of radiation cured epoxy resin–polyether sulfone blends as matrices for structural composites. *Polym. Degrad. Stab.* **2010**, *95*, 677–683. [CrossRef]

30. Lach, R.; Grellmann, W. Fracture mechanical properties. Thermosets and high performance composites. In *Mechanical and Thermomechanical Properties of Polymers. Landolt-Börnstein, Group VIII Advanced Materials and Technologies, Polymer Solids and Polymer Melts, New Series VIII/6A3*; Grellmann, W., Seidler, S., Eds.; Springer-Verlag: Berlin, Germany, 2014; pp. 423–492.

31. Janke, C.J.; Lomax, R.D.; Robitaille, S.; Duggan, S.; Serranzana, R.C.; Lopata, V.J. Improved epoxy resins cured by electron beam irradiation. In *A Materials and Processes Odyssey, Proceedings of 46th International SAMPE Symposium, Long Beach, CA, USA, 6–10 May 2001*; Repecka, L., Ed.; CRC Press: Boca Raton, FL, USA, 2001; pp. 2115–2126.

32. Serrano, E.; Martin, M.D.; Tercjak, A.; Pomposo, J.; Mecerreyes, D.; Mondragon, I. Nanostructured thermosetting systems from epoxidized styrene–butadiene block copolymers. *Macromol. Rapid Commun.* **2005**, *26*, 982–985. [CrossRef]

33. Ocando, C.; Tercjak, A.; Martin, M.D.; Serrano, E.; Ramos, J.A.; Corona-Galván, S.; Parellada, M.D.; Mondragon, I. Micro- and macrophase separation of thermosetting systems modified with epoxidized styrene-block-butadiene-block-styrene linear triblock copolymers and their influence on final mechanical properties. *Polym. Int.* **2008**, *57*, 1333–1342. [CrossRef]

34. Meng, Y.; Zhang, X.; Du, B.; Zhou, B.; Zhou, X.; Qi, G. Thermosets with core–shell nanodomain by incorporation of core crosslinked star polymer into epoxy resin. *Polymer* **2011**, *52*, 391–399. [CrossRef]

35. Khatiwada, S.P.; SarathChandran, C.; Lach, R.; Liebscher, M.; Saiter, J.M.; Thomas, S.; Heinrich, G.; Adhikari, R. Morphology and mechanical properties of star block copolymer modified epoxy resin blends. *Mater. Today: Proc.* **2017**, *4*, 5734–5742. [CrossRef]

36. Khatiwada, S.P.; Gohs, U.; Janke, A.; Jehnichen, D.; Heinrich, G.; Adhikari, R. Influence of electron beam irradiation on the morphology and mechanical properties of styrene/butadiene triblock copolymers. *Radiat. Phys. Chem.* **2018**, *152*, 56–62. [CrossRef]

37. Munteanu, S.B.; Vasile, C. Spectral and thermal characterization of styrene–butadiene copolymer with different architectures. *J. Optoelectron. Adv. Mater.* **2005**, *7*, 3135–3148.

38. George, S.M.; Puglia, D.; Kenny, J.M.; Parameswaranpillai, J.; Vijayan, P.P.; Pionteck, J.; Thomas, S. Volume shrinkage and rheological studies of epoxidised and unepoxidised poly(styrene-block-butadiene-block-styrene) triblock copolymer-modified epoxy resin–diamino diphenyl methane nanostructured blend systems. *Phys. Chem. Chem. Phys.* **2015**, *17*, 12760–12770. [CrossRef]

39. Pogany, G.A. Gamma relaxation in epoxy resins and related polymers. *Polymer* **1970**, *11*, 66–78. [CrossRef]

40. Sun, Y.; Zhang, Z.; Moon, K.S.; Wong, C.P. Glass transition and relaxation behavior of epoxy nanocomposites. *J. Polym. Sci. Part B: Polym. Phys.* **2004**, *42*, 3849–3858. [CrossRef]

41. Nogralo, F.F.; Llano-Ponte, R.; Mondragon, I. Dynamic and mechanical properties of epoxy networks obtained with PPO based amines/mPDA mixed curing agents. *Polymer* **1996**, *37*, 1589–1600. [CrossRef]

42. Young, C. Predicting practical properties of unfilled and filled adhesives from thermomechanical data. In *Adhesives and Consolidates for Conservation: Research and Applications, Proceedings of CCI Symposium*; Canadian Conservation Institute: Ottawa, ON, Canada, 2011; p. 20.

43. Vignoud, L.; David, L.; Sixou, B.; Vigier, G.; Stevenson, I. Effect of electron irradiation on the mechanical properties of DGEBA/DDM epoxy resins. *Nucl. Instrum. Methods Phys. Res. B* **2001**, *185*, 336–340. [CrossRef]

44. Ocando, C.; Tercjak, A.; Martín, M.D.; Ramos, J.A.; Campo, M.; Mondragon, I. Morphology development in thermosetting mixtures through the variation on chemical functionalization degree of poly(styrene-b-butadiene) diblock copolymer modifiers. Thermomechanical properties. *Macromolecules* **2009**, *42*, 6215–6224. [CrossRef]

45. Ruiz-Pérez, L.; Royston, G.J.; Fairclough, J.P.A.; Ryan, A.J. Toughening by nanostructure. *Polymer* **2008**, *49*, 4475–4488. [CrossRef]

46. Liu, J.; Thompson, Z.J.; Sue, H.J.; Bates, F.S.; Hillmyer, M.A.; Dettloff, M.; Jacob, G.; Verghese, N.; Pham, H. Toughening of epoxies with block copolymer micelles of wormlike morphology. *Macromolecules* **2010**, *43*, 7238–7243. [CrossRef]

47. Serrano, E.; Tercjak, A.; Ocando, C.; Larrañaga, M.; Perellada, M.D.; Corona-Galvan, S.; Mecerreyes, D.; Zafeiropoulos, N.E.; Stamm, M.; Mondragon, I. Curing behavior and final properties of nanostructured thermosetting systems modified with epoxidized styrene–butadiene linear diblock copolymers. *Macromol. Chem. Phys.* **2007**, *208*, 2281–2292. [CrossRef]

48. Bhandari, N.L.; Lach, R.; Grellmann, W.; Adhikari, R. Depth-dependent microindentationmicrohardness studies of different polymer nanocomposites. *Macromol. Symp.* **2012**, *315*, 44–51. [CrossRef]

49. Lach, R.; Michler, G.H.; Grellmann, W. Microstructure and indentation behaviour of polyhedral oligomeric silsesquioxanes-modified thermoplastic polyurethane nanocomposites. *Macromol. Mater. Eng.* **2010**, *295*, 484–491. [CrossRef]

50. Lange, F.F.; Radford, K.C. Fracture energy of an epoxy composite system. *J. Mater. Sci.* **1971**, *6*, 1197–1203. [CrossRef]

51. Sultan, J.N.; McGarry, F.J. Effect of rubber particle size on deformation mechanisms in glassy epoxy. *Polym. Eng. Sci.* **1973**, *13*, 29–34. [CrossRef]

52. Bagheri, R.; Pearson, R. Role of particle cavitation in rubber-toughened epoxies: II. Inter-particle distance. *Polymer* **2000**, *41*, 269–276. [CrossRef]

53. Ge, Z.; Zhang, W.; Huang, C.; Luo, Y. Study on epoxy resin toughened by epoxidizedhydroxy-terminated polybutadiene. *Mater.* **2018**, *11*, 932. [CrossRef] [PubMed]

54. Yi, F.; Yu, R.; Zheng, S.; Li, X. Nanostructured thermoset from epoxy poly (2,2,2-trifluro ethyl acrylate)-block-poly(glycidyl methacrylate) diblock copolymer: Demixing of reactive block and thermomechanical properties. *Polymer* **2011**, *52*, 5669–5680. [CrossRef]

materials

MDPI

Communication

Lignin as a Functional Green Coating on Carbon Fiber Surface to Improve Interfacial Adhesion in Carbon Fiber Reinforced Polymers

**László Szabó [1,*], Sari Imanishi [1], Fujie Tetsuo [1], Daisuke Hirose [1], Hisai Ueda [2],
Takayuki Tsukegi [2], Kazuaki Ninomiya [3] and Kenji Takahashi [1,*]**

[1]   Institute of Science and Engineering, Kanazawa University, Kakuma-machi, Kanazawa 920-1192, Japan;
     sarin0509@stu.kanazawa-u.ac.jp (S.I.); tfujie@p.kanazawa-u.ac.jp (F.T.); dhirose@se.kanazawa-u.ac.jp (D.H.)
[2]   Innovative Composite Center, Kanazawa Institute of Technology, 2-2 Yatsukaho, Hakusan 924-0838, Japan;
     h-ueda@neptune.kanazawa-it.ac.jp (H.U.); tsukegi@neptune.kanazawa-it.ac.jp (T.T.)
[3]   Institute for Frontier Science Initiative, Kanazawa University, Kakuma-machi, Kanazawa 920-1192, Japan;
     ninomiya@se.kanazawa-u.ac.jp
*   Correspondence: szabo-laszlo@se.kanazawa-u.ac.jp (L.S.); ktkenji@staff.kanazawa-u.ac.jp (K.T.);
     Tel.: +81-76-234-4828 (L.S. & K.T.)

Received: 30 November 2018; Accepted: 31 December 2018; Published: 6 January 2019

Abstract: While intensive efforts are made to prepare carbon fiber reinforced plastics from renewable sources, less emphasis is directed towards elaborating green approaches for carbon fiber surface modification to improve the interfacial adhesion in these composites. In this study, we covalently attach lignin, a renewable feedstock, to a graphitic surface for the first time. The covalent bond is established via aromatic anchoring groups with amine functions taking part in a nucleophilic displacement reaction with a tosylated lignin derivative. The successful grafting procedures were confirmed by cyclic voltammetry, X-ray photoelectron spectroscopy, and field emission scanning electron microscopy coupled with energy dispersive X-ray spectroscopy. Both fragmentation and microdroplet tests were conducted to evaluate the interfacial shear strength of lignin coated carbon fiber samples embedded in a green cellulose propionate matrix and in a commercially used epoxy resin. The microdroplet test showed ~27% and ~65% increases in interfacial shear strength for the epoxy and cellulose propionate matrix, respectively. For the epoxy matrix covalent bond, it is expected to form with lignin, while for the cellulosic matrix hydrogen bond formation might take place; furthermore, plastisizing effects are also considered. Our study opens the gates for utilizing lignin coating to improve the shear tolerance of innovative composites.

Keywords: Carbon fiber; epoxy composite; cellulose derivative; lignin; surface modification; interfacial adhesion

1. Introduction

The green energy policy not only involves the development of novel strategies for energy harvesting, but also places special emphasis on the improvement of the energy-efficiency of already existing technologies. Based on this principle, fuel-efficiency has been considered as a crucial requirement for creating a sustainable world. Therefore, legislative bodies are putting more and more challenging fuel consumption standards into force for manufacturers of the transportation sector all over the world [1,2]; these requirements can be met reasonably by including more and more lightweight materials into the structure of the vehicle [3]. As leading high-performance lightweight materials, carbon fiber reinforced polymers (CFRPs) caught, therefore, appreciable interest both in the industrial sector and in the scientific community. Based on a steady market growth with an annual growth rate above 10% since 2010 [4], it can be anticipated that CFRPs will indispensably shape our future. However, as petroleum-based energy intensive materials, carbon fiber reinforced polymers barely meet sustainability goals, and therefore, appreciable efforts are being made to fabricate green materials for future applications.

On one hand, a considerable amount of studies have been devoted to produce carbon fibers from renewable sources, such as lignin [5]. As a result, the mechanical strength of lignin-based carbon fibers (tensile strength lying between 0.6 and 1 GPa [5]) is not so far from the strength of carbon fibers available on the market (e.g., carbon fiber T300 from Toray has a tensile strength of ~3.5 GPa), and further improvement is expected in this field. On the other hand, special emphasis has been placed on changing the thermoset matrix (dominating the CFRPs market with ~70% contribution [4]) to environmentally more benign thermoplastic polymers [6]. Furthermore, cellulose-based carbon fiber reinforced polymers are also gaining attention [7,8], and advances in this field open the gates for the preparation of fully biomass-derived CFRPs.

Nevertheless, due to the relatively inert and hydrophobic nature of the carbon fiber surface, the interfacial adhesion between the matrix and the fiber in the composite is notoriously low, leading to limitations in mechanical performance that highly depends on this factor [9,10]. Therefore, many studies have been focused on the surface modification of carbon fibers, giving rise to a wide range of successful modification techniques to increase interfacial shear strength (IFSS) [9–23]. At the same time, most of the studies hardly take sustainability issues and green chemistry principles into account as the synthetic steps usually involve aggressive chemicals to bias the graphitic surface.

Keeping in mind green chemistry principles [24], and also inspired by recent achievements in our laboratory in respect to biomass valorization, we intended to utilize lignin as a functional green coating on carbon fiber surface to improve interfacial properties of a cellulose-based (cellulose propionate, thermoplastic), and a commercially used epoxy (thermosetting) composite. While an epoxy resin is not considered as a green matrix, we intended to include this type of polymer in our study based on advances in making epoxy resins from renewable sources [25–27]. Furthermore, there is an obvious need for improving the mechanical properties of these composites [9,10,28]. Cellulose propionate was chosen as a representative of green biomass-derived renewable plastics, this cellulose ester has superior characteristics over other cellulose derivatives available on the market (good processability at high temperature and high tensile strength due to its high molecular weight). Lignin is the most abundant renewable feedstock for aromatic compounds, and it is available in large quantities as a byproduct of the pulping industry [29,30]. A covalent lignin coating is expected to improve interfacial shear strength in cellulose-based composites via hydrogen bonding interactions (between free OH groups, and due to -OH----O=< interactions), and by acting as a plasticizer at the interphase [31], such a mechanical dynamical effect was shown to enhance the shear tolerance of CFRPs [32]. In an epoxy composite, covalent bond formation can be envisaged as a result of the reaction between epoxy groups of the resin and the OH groups of lignin, and plasticizing effects might also have an influence on IFSS. Our study is aimed at testing these theories by covalently binding lignin to the surface of carbon fiber and investigating the mechanical performance at the interphase.

2. Materials and Methods

2.1. Materials

Cellulose propionate ($M_{\mathrm{w}} \approx 200,000$ g mol^{-1} according to the manufacturer) was supplied by Scientific Polymer Products, Inc. (Ontario, NY, USA), the degree of substitution was determined to be 2.76 in our previous publication (92% of hydroxyl groups are substituted) [7,8]. The matrix has a tensile strength of ~80 MPa and a glass transition temperature (T_g) of ~132 °C (Shimadzu DSC-60A Plus, Kyoto, Japan). Bisphenol type epoxy matrix was prepared using West System® 105 Epoxy Resin and West System® 206 Slow Hardener with a 1:5 (v/v) ratio according to the recommendations of the producer (Gougeon Brothers, Inc., Bay City, MI, USA). The epoxy matrix has a tensile strength of ~50 MPa and a glass transition temperature of ~59 °C.

Kraft lignin (low sulfonate content; $M_w \approx 10,000$ g mol^{-1} according to the manufacturer; Lot # 04414PEV) was obtained from Sigma Aldrich (St. Louis, MO, USA). The OH content was determined using ^{31}P-NMR measurement (JEOL 600 MHz FT-NMR spectrometer, JEOL Ltd., Tokyo, Japan) according to the procedure reported by Granata and Argyropoulos [33]. The free hydroxyl group content was calculated as follows: OH$_{aliphatic}$ = 1.65 mmol g^{-1}, OH$_{phenolic}$ = 2.12 mmol g^{-1}, OH$_{carboxylic}$ = 0.59 mmol g^{-1}, OH$_{total}$ = 4.36 mmol g^{-1} (^{31}P-NMR spectrum is shown in Figure S1).

Carbon fiber T700SC-12000-50C (~7 μm fiber diameter; ~4.9 GPa tensile strength) was purchased from Toray Industries (Tokyo, Japan). The fibers have nominally ~1% sizing agent on the surface, therefore, in order to access the graphitic surface for the grafting reactions, a cleaning procedure was applied [7,8], the removal of the commercial coating was evidenced in our previous report [7].

3-Ferrocenylpropionic anhydride, anhydrous dimethyl sulfoxide, anhydrous pyridine, and 4-[(*N*-Boc)aminomethyl]aniline were provided by Sigma-Aldrich (St. Louis, MO, USA). Acetonitrile and hydrochloric acid were from Naclai Tesque, Inc. (Kyoto, Japan). *Tert*-butyl nitrite (isoamyl nitrite), *ortho*-dichlorobenzene, 1,4-dioxane, *p*-toluenesulfonyl chloride (tosyl chloride, TsCl), 4-dimethylaminopyridine (DMAP), and tetrabutylammonium hexafluorophosphate were purchased from Tokyo Chemical Industry Co., Ltd. (Tokyo, Japan). All the other solvents were obtained from Kanto Chemical (Tokyo, Japan).

2.2. Methods

2.2.1. Synthesizing Lignin Derivatives

One gram kraft lignin (4.36 mmol OH content) was dispersed in 100 mL pyridine. Under continuous stirring, 0.83 g tosyl chloride (~1 equivalent in respect to OH content) dissolved in 20 mL pyridine was added dropwise to the latter solution in 30 min. The reaction was conducted at 8 °C for 24 h, and the product was obtained by pouring the reaction mixture into cold ethanol. The resulting product was washed with acetone and dried in a vacuum oven at 50 °C for 3 days. The afforded tosylated lignin (Figure 1) showed good solubility in pyridine and DMSO. Since the product was not soluble in the conventional solvent system of the ^{31}P-NMR measurement [33], the OH content and successful tosylation was confirmed by elemental analysis (Vario ELCUBE and Vario EL III, Elementar Analysensysteme GmbH, Langenselbold, Germany). From the elemental analysis results, we calculated that the OTs content amounts to 0.26 mmol g^{-1}.

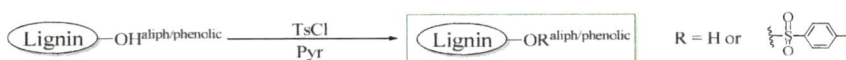

Figure 1. Reaction scheme for synthesizing tosylated lignin. Abbreviations: –OH^{aliph}—aliphatic hydroxyl groups in lignin (1.65 mmol g^{-1} as determined by ^{31}P-NMR measurement); –OH^{phenolic}—phenolic hydroxyl groups in lignin (2.12 mmol g^{-1} as determined by ^{31}P-NMR measurement); TsCl—tosyl chloride; and Pyr—pyridine.

2.2.2. Carbon Fiber Surface Modification

Grafting 4-(Aminomethyl)Benzene Functions onto the Carbon Fiber Surface

4-(Aminomethyl)benzene functions were deposited on the carbon fiber surface using 4-[(*N*-Boc)aminomethyl]aniline as the starting molecule (Figure 2). The diazonium salt (29.2 μmol/mg fibers) was in situ prepared with isoamyl nitrite (58.4 μmol/mg fibers) in *ortho*-dichlorobenzene/acetonitrile (2:1 volume fraction) mixed solvent system, and decomposed by heating (50 °C, 24 h) in the presence of carbon fibers. After the reaction, the fibers were washed thoroughly with dichloromethane, deionized water, and acetone. The deprotection step involved the immersion of the fibers in 2 M anhydrous HCl solution (in 1,4-dioxane), followed by a neutralization step with 2 M aqueous NaOH solution. The functionalized fibers were washed again with dichloromethane, deionized water, and acetone, and dried in a vacuum oven at 50 °C for 24 h. This free radical-mediated grafting procedure was shown to be a benign means of functionalization not affecting key single carbon fiber mechanical parameters [17]. A detailed description of the experimental procedure is given in our previous publication [7].

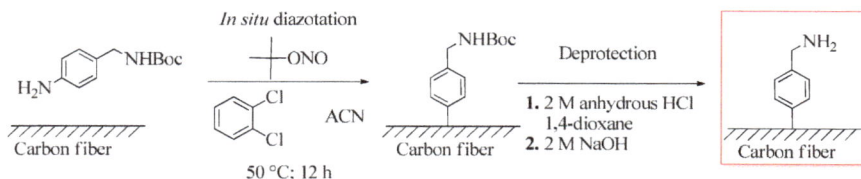

Figure 2. Reaction scheme for grafting 4-(aminomethyl)benzene functions onto the carbon fiber surface. Abbreviations: Boc—*tert*-butyloxycarbonyl protecting group; and ACN—acetonitrile.

Derivatization of 4-(Aminomethyl)Benzene Functions for Determining the Grafting Density

The deposited 4-(aminomethyl)benzene functions were further derivatized in order to be able to precisely quantify the amount of grafted structures on the surface. For this purpose, an electrochemical probe (Figure 3) was covalently linked to these structures according to the following procedure. Briefly, 25 mg functionalized fibers were immersed in 50 mL pyridine and reacted with 0.36 g 3-ferrocenylpropionic anhydride (0.73 mmol) in the presence of 0.09 g DMAP (0.73 mmol) at 80 °C for 24 h. The fibers thus obtained were washed with dichloromethane, deionized water, and acetone, and dried in a vacuum oven at 50 °C for 24 h.

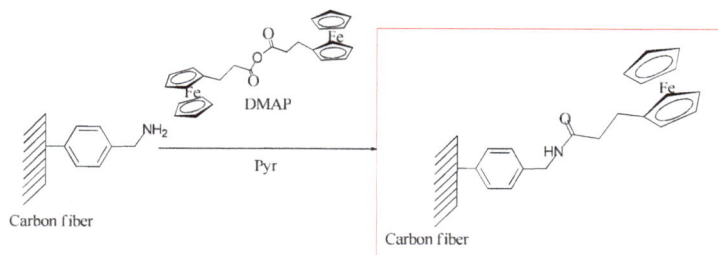

Figure 3. Reaction scheme for binding electrochemically active structure to the 4-(aminomethyl)benzene functions. Abbreviations: DMAP—4-dimethylaminopyridine; and Pyr—pyridine.

Grafting Lignin Derivative onto the Carbon Fiber Surface

Five gram tosylated lignin was dissolved in 50 mL DMSO and added to a solution containing 0.1 mL triethylamine (0.73 mmol) and 25 mg 4-(aminomethyl)benzene functionalized carbon fiber in 20 mL DMSO (Figure 4). The reaction was conducted at 100 °C for 24 h. The fibers were washed, as previously stated, and dried at 50 °C for 24 h in a vacuum oven.

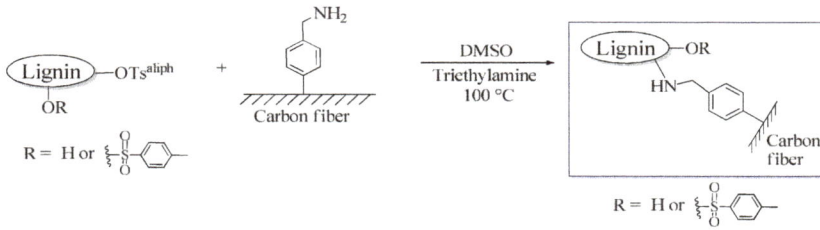

Figure 4. Immobilizing lignin derivative on the carbon fiber surface. Abbreviations: –OTs^{aliph}—tosylated aliphatic hydroxyl groups; and DMSO—dimethyl sulfoxide.

2.2.3. Mechanical Tests

Fragmentation Test

Fragmentation test was conducted according to our previous studies using cellulose propionate as polymer matrix [7,8]. A hydraulic hot-press machine (Type MH-10, Imoto Machinery Co., Kyoto, Japan) was used to prepare polymer films and single fiber composites (10 specimens from each type of carbon fibers). Plastic film with a thickness of ~0.15 mm was prepared by firstly placing 1 g of cellulose propionate between the clams of the hot press machine at 203 °C for 7 min, followed by pressing with 45 kN load for 3 min. The film was removed after the temperature dropped below 193 °C. Single fibers were placed between two plastic films and the same hot pressing process was applied. Thereafter, standard dumbbell specimens (the exact dimensions are given in our previous work [7]) were cut from the single fiber composite films. The specimens were elongated longitudinally with a Shimadzu Autograph AG-X Plus 5 kN tensile tester (Kyoto, Japan) applying a crosshead speed of 0.5 mm min^{-1}. The fragmentation process was monitored with a high resolution digital camera (N.O.W.-D2X3Z-KSH, Nihonkouki, Aoki, Japan), and the fragment size after saturation (no shorter fragment size observed) was determined using an AUSB-K version 14.4 program (Nihonkouki, Aoki, Japan) with the help of an objective micrometer (Shibuya Optical Co., Ltd., Wako, Japan). For each specimen, the elongation at break exceeded 20 mm, and the saturation point could be clearly achieved. The Kelly-Tyson model [34,35] is applied to calculate the apparent interfacial shear strength according to Equations (1)–(3):

$$\tau = (\sigma_{fu} \times d)/(2 \cdot l_c) \tag{1}$$

$$l_c = 4/5 \cdot l \tag{2}$$

$$\sigma_{fu} = \sigma_1 \times (l_1/l_c)^{(1/m)} \tag{3}$$

where σ_{fu} is the fiber tensile strength at critical length, d is the fiber diameter and l_c denotes critical length, which is defined according to Equation (2) involving the average fragment size (l). The σ_{fu} can be calculated after σ_1 is determined from single fiber tensile strength experiments conducted at l_1 gauge length [35,36]. The Weibull modulus (m) represents the data spread of single fiber tensile strength experiments.

Microdroplet Test

Microdroplet experiment was conducted on an HM410 equipment designed for the evaluation of fiber/resin composite interface properties, manufactured by Tohei Sangyo Co., Ltd. (Tokyo, Japan). Droplets on single fibers were prepared with diameters between 80–100 μm (droplets with sizes outside this range were not measured), the experimental setup is shown in Figure S2. For preparation of droplets from the epoxy matrix, epoxy resin and hardener was mixed in a 1:5 (*v/v*) ratio according to the recommendations of the producer. Single fibers fixed on a metal frame were immersed in this mixture, and droplets were formed due to the Rayleigh instability. The samples were kept in a drying oven at 80 °C for 3 days. In case of cellulose propionate, the polymer was melted on a TJA-550 hot plate (AS ONE Corporation, Osaka, Japan) at 235 °C (5 min), and then droplets were formed at 250 °C (3 min). These samples were kept at room temperature. The microdroplet test was conducted at room temperature with a pull-out speed of 0.06 mm min^{-1} and 1 N load cell was applied. The interfacial shear strength was calculated for the maximum load (F) according to Equation (4).

$$\tau = F/\pi dL \tag{4}$$

where d is the fiber diameter and L is the embedded length (droplet length).

Fifty specimens were prepared for the microdroplet test from each type of carbon fibers. Significance analysis was performed using multiple t-test, assuming that the data represents a population with equal variance and $\alpha = 0.05$.

2.2.4. Surface Analysis

X-ray photoelectron spectroscopy (XPS) was conducted on a Thermo Scientific K-Alpha X-ray Photoelectron Spectrometer System equipped with an Al K_{α} monochromated X-ray source (1486.6 eV) having a power of 36 W (12 kV × 3 mA) (Waltham, MA, USA). The analysis spot size was set to 400 μm. The binding energy scale was calibrated using the hydrocarbon C1s peak at 285.0 eV. High-resolution C1s, O1s, N1s, and S2p spectra were recorded at 20 eV pass energy with 0.1 eV resolutions. The spectra were analyzed with a Thermo Scientific Avantage Software version 5.89 (Waltham, MA, USA). Background correction was executed using a built-in Smart algorithm and peak fitting was performed with the Powell method applying a Gauss-Lorentz Mix algorithm.

Field emission scanning electron microscopy (FE-SEM) images were obtained with a JSM-7610F system (JEOL, Tokyo, Japan), which was equipped with a JEOL EX-230**BU EX-37001 Energy Dispersive X-Ray Analyzer (JEOL, Tokyo, Japan) allowing to acquire EDX spectra and perform chemical mapping experiments. The FE-SEM analysis was conducted with a working distance of 8 mm applying an accelerating voltage of 15 kV.

SEM images were recorded on a Hitachi S4500 system (accelerating voltage of 15 kV, Hitachi, Ltd., Tokyo, Japan). Samples were coated with Au/Pd layer for 40 s using a Hitachi E1030 ion sputter (Hitachi, Ltd., Tokyo, Japan).

Cyclic voltammetry experiments were performed using an ALS/CHInstruments Electrochemical Analyzer Model 1200A potentiostat connected to an SVC3 voltammetry cell (ALS Co., Ltd., Tokyo, Japan). The three electrode system consisted of a platinum counter electrode (ALS Co., Ltd., Tokyo, Japan), an Ag/Ag$^+$ non-aqueous reference electrode (RE-7, ALS Co., Ltd., Tokyo, Japan), and carbon fiber was used as working electrode attached to the terminal with a copper tape. The cell containing 0.1 M tetrabutylammonium hexafluorophosphate in acetonitrile (supporting electrolyte) was cycled between −0.4 V and 0.4 V, with a scan rate of 0.1 V s^{-1}. The area under the reduction and oxidation peak was integrated to obtain the charge (in Coulomb) transferred per 1 s, and by dividing this value with the scan rate we could calculate the total charge transferred (Q). The surface coverage can be determined by the following equation in case of the ferrocene/ferrocenium couple (one electron transfer, m denotes the mass of the functionalized carbon fiber working electrode) [21]:

$$\text{Surface coverage (mol mg}^{-1}) = Q/96485m \tag{5}$$

The cyclic voltammetry experiment was conducted with three samples and both the reduction and oxidation peaks were taken into account when calculating (from the first cycles) the average surface coverage and the associated deviation.

3. Results and Discussion

3.1. Grafting Lignin onto the Carbon Fiber Surface

Lignin was bound covalently to the graphitic surface of carbon fibers according to Figures 1, 2 and 4. Our synthetic strategy involved in situ grafting of a 4-(aminomethyl)benzene moiety onto the surface using diazonium species (Figure 1) followed by a simple nucleophilic displacement reaction with tosylated lignin (Figure 4). The mechanism of functionalization via diazonium species can be explained in terms of free radical processes. Diazonium species can yield aryl radicals upon heating, these radicals can add to double bonds or to aromatic systems [37]. This type of functionalization for an extended aromatic graphitic system was reported first for carbon nanotubes [38], and later adopted to carbon fiber surface chemistry [17]. The further S_N2 type alkylation of an amino group via the tosylated derivative of a biopolymer is a well-described reaction in the literature [39].

3.1.1. Quantifying Grafted 4-(Aminomethyl)Benzene Functions on the Surface

The presence of a 4-(aminomethyl)benzene moiety, and thereby the validation of the first step of our synthetic strategy (Figure 1), has been proved in our previous study by XPS analysis [7]. However, the amount of grafted structures on the surface has not been quantified. In order to be able to build other, environmentally more benign synthetic paths on our methodology in the future, we intended to take this issue under scrutiny. For this purpose, the originally grafted 4-(aminomethyl)benzene structure was modified with an electrochemical probe (ferrocene/ferrocenium redox couple), and cyclic voltammetry experiment was performed. The cyclic voltammogram thus obtained is shown in the Supplementary Material (Figure S3), and from these data, the surface coverage was determined to be $2.96 \pm 1.6 \times 10^{14}$ molecules mg^{-1}. The grafting density in our study is higher than that reported by Li et al. [12] (8.15×10^{12} molecules mg^{-1} for grafted 4-aminobenzene structures; they used different reaction conditions, e.g., isopentyl nitrite as reactant and water as solvent), and somewhat lower than that reported for electrochemical grafting of a diazonium salt ($\sim1.2 \times 10^{15}$ molecules mg^{-1} for grafted phenylacetylene structures) [21].

3.1.2. Surface Characterization—Experimental Evidence for Lignin Coating

X-ray photoelectron spectroscopy experiments were performed to confirm successful functionalization of the surface with lignin, and thereby validating our synthetic procedure (Figures 1, 2 and 4). The high-resolution C1s and O1s spectra are shown in Figure 5a,b, respectively (survey spectrum is shown in Figure S4).

The high resolution C1s spectrum of the unfunctionalized carbon fiber surface (Figure 5a Inset) exhibits a broad and characterless peak assigned to localized (amorphous) as well as delocalized alternant hydrocarbons [40,41]. Furthermore, a broad peak arises at higher binding energies representing defect sites with –COOH functions, and π-π* shake-up peaks are also present around this region [41]. At the same time, after the functionalization procedure (Figures 1, 2 and 4), the high-resolution C1s spectrum (Figure 5a) indicates the presence of additional peaks that can be attributed to C–OH (286.22 eV), C=O (287.41 eV), and COOH (288.88 eV) moieties deposited on the surface [41,42]. The relatively large C–O peak suggests appreciable contribution of C–OH functions to the C1s spectrum, and therefore, an enhanced amount of these functions on the surface. Such a carbon-oxygen bond distribution profile clearly indicates the presence of lignin. To gain further knowledge about surface chemical composition, the O1s spectrum was also recorded (Figure 5b). Compared to the O1s spectrum of the unfunctionalized sample (Figure 5b Inset), the spectrum of the functionalized sample can be resolved to further components indicating C–O and C=O bonds on the surface [41], which have a similar contribution to that shown in the C1s spectrum (Figure 5a). This result additionally confirms a successful grafting procedure leading to a lignin-coated surface. In addition, the S2p spectrum of the functionalized sample (Figure S5) shows a small peak localized around the binding energies, characteristic for tosylates ($\sim$168–169 keV) [43], and the weak intensity compared to the unfunctionalized sample (Figure S5 Inset) suggests that most of the tosyl groups took place in the nucleophilic displacement reaction. We have also acquired the high resolution N1s spectra of the unfunctionalized fibers together with the 4-(aminomethyl)benzene and lignin functionalized samples (Figure S6). The N1s spectra indicate very small amounts of N on the surface of these materials. Due to the spectral similarities and very low intensities compared to the spectral noise, firm conclusions cannot be drawn from the N1s spectra.

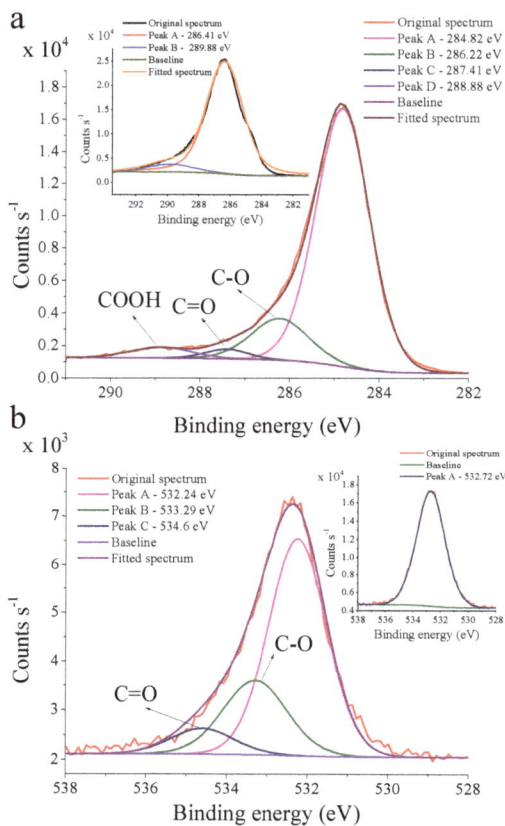

Figure 5. (**a**) C1s and (**b**) O1s spectra of carbon fiber sample with lignin coating, insets show the C1s (**a** Inset) and O1s (**b** Inset) spectra recorded on the clean carbon fiber surface before the functionalization procedure.

To further analyze the bulk structure of the fibers and the near-surface layers, energy dispersive X-ray analysis was also performed (the penetration depth reaches the micrometer scale, note that the carbon fiber diameter is ~7 μm). Elemental compositions obtained from EDX analysis are shown in the Supplementary Material (Table S1). Compared to the control sample (no sizing agent, unfunctionalized), a reasonable increase in the oxygen content can be noticed for the lignin-coated sample. Furthermore, the nitrogen content remains similar to the control sample. Based on the results of the chemical mapping (Figure S7), the fiber structure does not suffer damage during our synthetic procedure, which is very important for keeping crucial carbon fiber mechanical parameters that are eventually imparted to the final composite.

The FE-SEM (Figure S8) and SEM (Figure S9) images indicate that the control fiber surface exhibits enhanced surface roughness compared to the lignin functionalized sample and the sample with a sizing agent on the surface (carbon fiber as received). The FE-SEM and SEM images point out the presence of smooth lignin coating on the surface.

3.2. Mechanical Tests

3.2.1. Fragmentation Test

Dumbbell specimens for the fragmentation test were prepared using injection molding technique according to our previous studies [7,8]. As a green thermoplastic matrix, cellulose propionate was chosen for these experiments due to its superior properties among cellulose-based matrices available on the market (high molecular weight, good processability [7,8]). Epoxy resin was not used for the fragmentation test (difficulties in preparing single fiber composite for the test), but it was involved in the microdroplet experiment. The results of the fragmentation test are shown in Figure 6. The IFSS increases ~28% for 4-(aminomethyl)benzene functionalized single fiber composites compared to the unfunctionalized sample (no sizing). This phenomenon was attributed, in our previous study, to hydrogen bonding interactions between the amino groups on the surface of fibers and oxygen atoms of the cellulose propionate matrix [7]. A similarly high IFSS value was also obtained for the lignin coated carbon fiber sample. According to our theory, in this case, hydrogen bonding interactions and mechanical dynamical effects (plasticizing effect [31]) can take place, improving interfacial shear strength. The SEM images of the fracture surfaces (Figure S10) are also in line with the IFSS values since the hole surrounding the fibers after delamination is smaller as the IFSS increases, indicating stronger fiber-matrix interactions.

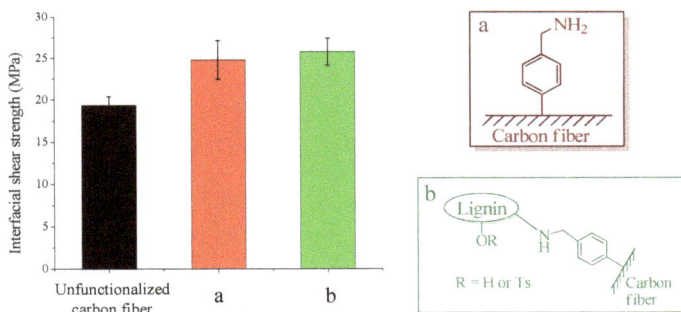

Figure 6. Interfacial shear strength determined indirectly using fragmentation test. The test was performed on composites prepared using cellulose propionate as matrix and unfunctionalized carbon fiber, (**a**) 4-(aminomethyl)benzene functionalized carbon fiber, or (**b**) carbon fiber containing lignin on the surface.

3.2.2. Microdroplet Test

The microdroplet test is considered to be a direct method to evaluate the interfacial shear strength; the pull-out force is monitored in real-time during the experiment. The results of the microdroplet test are shown in Figure 7.

A decrease in IFSS is noticed for 4-(aminomethyl)benzene functionalized carbon fiber samples embedded in the epoxy matrix (compared to the unfunctionalized sample). This decrease is quite surprising since covalent bond is expected to form as a result of the reaction between the epoxy groups of the matrix and the amine functions on the surface. Such a covalent interaction between the matrix and fiber was shown to considerably increase the interfacial shear strength [20,21]. We rationalized our finding in light of the increased reactivity of 4-(aminomethyl)benzene compared to aniline (previous studies [20]) towards epoxy reagents [44]. There is a competition between surface-grafted amine functions and the amine hardener for reacting with the epoxy groups, which might lead to a less interconnected, and thus weaker, epoxy matrix near the interface. The IFSS shows appreciable increase when lignin is present on the carbon fiber surface (~27% increase compared to the unfunctionalized sample), and in this case, a covalent bond can be formed involving the hydroxyl groups of lignin and epoxy groups of the matrix.

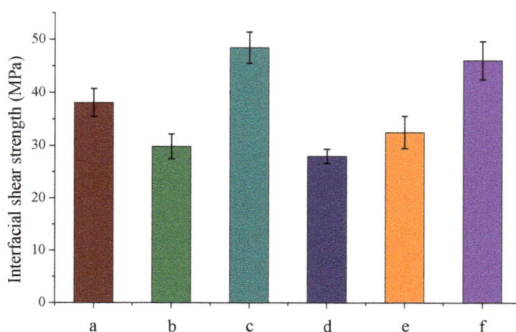

Figure 7. Results of the microdroplet experiments. (**a**) Epoxy resin—unfunctionalized carbon fiber (CF), (**b**) epoxy resin—4-(aminomethyl)benzene functionalized CF, (**c**) epoxy resin—lignin functionalized CF, (**d**) cellulose propionate—unfunctionalized CF, (**e**) cellulose propionate—4-(aminomethyl)benzene functionalized CF, and (**f**) cellulose propionate—lignin functionalized CF.

When cellulose propionate is used as the matrix, the IFSS value increases as 4-(aminomethyl)benzene functions are deposited on the carbon fiber surface (~16% increase) in line with the fragmentation test (Figure 6). However, a relatively large increase in IFSS (~65% increase) is experienced for the lignin functionalized carbon fiber sample compared to the fragmentation test. A similarly large increase was obtained previously for a thermoplastic polypropylene matrix and ionic liquid sizing agent, the latter acting as a plasticizer at the interface [32]. We assume that our result cannot be only discussed in terms of hydrogen bonding interactions, but plasticizing effects might also take place. Previous reports indicate that lignin might exert a plasticizing effect for cellulose esters (glass transition temperature decreases as lignin is added to the system) [31]. The difference found between the IFSS values of the fragmentation test and the microdroplet test is attributed to the inherently different experimental approaches. While during the microdroplet test the load is placed on a small resin droplet on the fiber, during the fragmentation test, the sample is elongated using an autograph and the load is transferred to the fiber through a large amount of the matrix, in the latter process, therefore, the load transfer is less efficient. The microdroplet test should be considered to be more relevant as it is a direct measure of IFSS. In order to substantiate and visualize the results of the microdroplet test, the fracture surfaces were monitored after debonding (Figure 8).

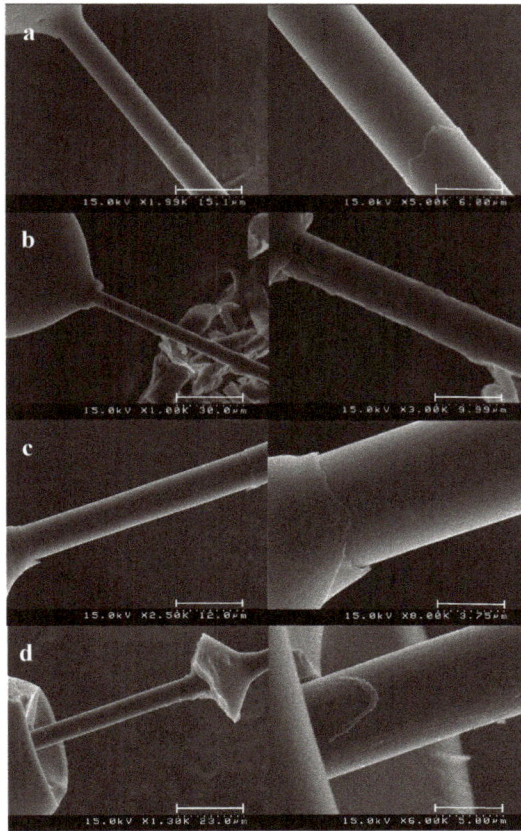

Figure 8. Fracture surfaces recorded after microdroplet test. (**a**) Epoxy matrix with unfunctionalized fiber, (**b**) epoxy matrix with lignin coated fiber, (**c**) cellulose propionate matrix with unfunctionalized fiber, and (**d**) cellulose propionate with lignin functionalized fiber.

In case of the unfunctionalized fibers (Figure 8a,c), both for the epoxy resin and for the cellulose propionate matrix, the delamination is smooth after reaching the maximum load during the microdroplet test, and furthermore, there is no remaining polymer on the surface indicating poor interfacial adhesion. When 4-(aminomethyl)benzene functions are present on the surface, however, some matrix remains on the fibers as expected (Figure S11a,b, note that the lower IFSS value for the epoxy resin is attributed to the formation of a less interconnected polymer network near the interface). In case of the epoxy resin droplet when the fibers are coated with lignin (Figure 8b), an appreciable amount of matrix remains on the surface, and matrix failure occurs indicating very strong interfacial interactions. Furthermore, for cellulose propionate matrix, a drastic failure takes place when lignin is deposited on the carbon fiber surface (Figure 8d), again the bulk matrix deteriorates before the interfacial layer and considerable amount of polymer is observable on the surface. Our results therefore indicate that, practically, the adhesion at the interface is maximized as the interfacial layer can resist more load than the matrix itself.

The effect of sizing agent on the interfacial shear strength for cellulose propionate matrix has been investigated in our previous study using the fragmentation test [7]. As the sizing agent was removed from the surface, the interfacial shear strength slightly increased (~10%). We attributed this increase to the unfavorable interactions between polar groups of the sizing agent and the hydrophilic alkyl chains of cellulose propionate. When carbon fiber is coated with lignin, an ~46% increase in interfacial shear strength can be noted compared to the fiber with sizing agent. At the same time, it appears that the presence/absence of sizing agent has only slight effect on the interfacial shear strength for an epoxy composite based on a study that used the same type of fibers and microdroplet test [45].

3.3. Practical Considerations

When carbon fiber surface chemistries are applied, it is crucial to avoid any damage to the fiber structure and thus keep single fiber strength. The first step of the functionalization procedure (grafting aromatic structure onto the carbon fiber surface) is known to be a nondestructive process, not impairing single fiber strength [17]. Grafting lignin onto this structure is thought to be still a benign process since there was no mechanical damage observed on the fiber.

It should be noted that we placed special emphasis on using a synthetic pathway that can be followed easily by surface analytical techniques (e.g., cyclic voltammetry, XPS spectroscopy, and FE-SEM EDX analysis). Especially because of the heterogeneity of carbon fiber surface, detecting functional groups on the surface is difficult [17,46]. Based on our results, an environmentally more benign synthetic strategy is planned to be applied and the process will be scaled-up to prepare large amount of carbon fiber reinforced composites.

4. Conclusions

Recent advances achieved in making carbon fibers from renewable sources, together with the successful preparation of cellulose-based carbon fiber reinforced composites, open the gates for the creation of fully bio-based high-performance composites. Nevertheless, the carbon fiber surface needs to be modified to improve the interfacial adhesion, these techniques usually involve environmentally less benign chemistries. In this study, we intended to use lignin as a functional green material to improve interfacial properties of epoxy and cellulose-based composites.

We found that covalent attachment of lignin to the carbon fiber surface leads to very strong interfacial interactions for epoxy and cellulose propionate matrices. Based on microdroplet experiments, in these composites matrixes, failure occurs before interfacial delamination could take place. Therefore, the mechanical tests suggest that covalent interactions (epoxy matrix), hydrogen bonding (cellulose propionate matrix), and plasticizing effects (cellulose propionate and epoxy matrix) might lead to improved shear tolerance materials when lignin is present on carbon fiber surface.

Supplementary Materials: The following are available online at http://www.mdpi.com/1996-1944/12/1/159/s1. Figure S1: Quantitative ^{31}P-NMR spectrum for determining OH content of kraft lignin sample (as received from Sigma Aldrich) according to Granata and Argyropoulos [31], Table S1: Elemental composition of the bulk and near-surface layer of carbon fiber samples, Figure S2: Microdroplet experimental setup, Figure S3: Cyclic voltammetry results to determine grafting density of 4-(aminomethyl)benzene structures modified with the ferrocene/ferrocenium couple (note that the cyclic voltammograms were obtained using different amount of samples), Figure S4: Survey spectrum recorded for carbon fiber sample containing lignin on the surface, Figure S5: High-resolution S2p spectra of carbon fiber sample containing lignin on the surface with inset displaying the unfunctionalized sample, Figure S6: High resolution N1s spectra of (a) unfunctionalized carbon fiber, (b) 4-(aminomethyl)benzene functionalized carbon fiber and (c) lignin functionalized samples, Figure S7: Chemical mapping experiment for carbon fiber sample containing lignin on the surface, Figure S8: FE-SEM images of (a) control carbon fiber (unfunctionalized), (b) carbon fiber with sizing agent (as received) and (c) carbon fiber containing lignin on the surface, Figure S9: SEM images of (a) control carbon fiber (unfunctionalized), (b) carbon fiber with sizing agent (as received) and (c) carbon fiber containing lignin on the surface, Figure S10: SEM images of fracture surfaces after the fragmentation test for (a) unfunctionalized carbon fiber sample, (b) carbon fiber functionalized with 4-(aminomethyl)benzene and (c) containing lignin coating on the surface, Figure S11: Fracture surfaces recorded after microdroplet test. 4-(Aminomethyl)benzene functionalized carbon fiber samples with (a) epoxy and (b) cellulose propionate matrix.

Author Contributions: Conceptualization, L.S. and K.T.; methodology, L.S., S.I., H.U., T.T.; formal analysis, L.S.; investigation, S.I. and L.S.; resources, F.T.; writing—original draft preparation, L.S.; writing—review and editing, K.T., K.N.; T.T., H.U. and D.H.; supervision, K.T., K.N., T.T., H.U. and D.H.; project administration, K.T. and K.N.; funding acquisition, K.T. and K.N.

Funding: This research was funded by the COI program: "Construction of next-generation infrastructure using innovative materials—Realization of a safe and secure society that can coexist with the Earth for centuries" supported by the Ministry of Education, Culture, Sports, Science and Technology (MEXT) and the Japan Science and Technology Agency (JST). Our study was also supported in part by the Advanced Low Carbon Technology Research and Development Program (ALCA) and the Cross-ministerial Strategic Innovation Promotion Program (SIP) of the JST.

Conflicts of Interest: The authors declare no conflict of interest.

References

1. U.S. Environmental Protection Agency. 2017 and later model year light-duty vehicle greenhouse gas emissions and corporate average fuel economy standards. *Fed. Reg.* **2012**, *77*, 62623–63200.

2. Regulation of the European Parliament and of the Council. *Setting Emission Performance Standards for New Passenger Cars and for New Light Commercial Vehicles as Part of the Union's Integrated Approach to Reduce CO2 Emissions from Light-Duty Vehicles and Amending Regulation (EC) No 715/2007*; European Parliament and of the Council: Brussels, Belgium, 2017.

3. Kyono, T. Life cycle assessment of carbon fiber-reinforced plastic. In *High-Performance and Specialty Fibers*; The Society of Fiber Science and Technology, Japan, Ed.; Springer: Tokyo, Japan, 2016; pp. 355–361. ISBN 978-4-431-55202-4.

4. Witten, E.; Sauer, M.; Kühnel, M. *Composites Market Report 2017. Market Developments, Trends, Outlook and Challenges*; Industrievereinigung Verstärkte Kunststoffe e.V. (Federation of Reinforced Plastics): Frankfurt, Germany, 2017.

5. Fang, W.; Yang, S.; Wang, X.-L.; Yuan, T.-Q.; Sun, R.-C. Manufacture and application of lignin-based carbon fibers (LCFs) and lignin-based carbon nanofibers (LCNFs). *Green Chem.* **2017**, *19*, 1794–1827. [CrossRef]

6. Yao, S.-S.; Jin, F.-L.; Rhee, K.Y.; Hui, D.; Park, S.-J. Recent advances in carbon-fiber-reinforced thermoplastic composites: A review. *Compos. Part B Eng.* **2018**, *142*, 241–250. [CrossRef]

7. Szabó, L.; Imanishi, S.; Kawashima, N.; Hoshino, R.; Takada, K.; Hirose, D.; Tsukegi, T.; Ninomiya, K.; Takahashi, K. Carbon fiber reinforced cellulose-based polymers: Intensifying interfacial adhesion between the fibre and the matrix. *RSC Adv.* **2018**, *8*, 22729–22736. [CrossRef]

8. Szabó, L.; Imanishi, S.; Kawashima, N.; Hoshino, R.; Hirose, D.; Tsukegi, T.; Ninomiya, K.; Takahashi, K. Interphase engineering of a cellulose-based carbon fiber reinforced composite by applying click chemistry. *Chem. Open* **2018**, *7*, 720–729. [CrossRef] [PubMed]

9. Sharma, M.; Gao, S.; Mäder, E.; Sharma, H.; Wei, L.Y.; Bijwe, J. Carbon fiber surfaces and composite interphases. *Compos. Sci. Technol.* **2014**, *102*, 35–50. [CrossRef]

10. Krager-Kocsis, J.; Mahmood, H.; Pegoretti, A. Recent advances in fiber/matrix interphase engineering for polymer composites. *Prog. Mater. Sci.* **2015**, *73*, 1–43. [CrossRef]

11. Wang, Y.; Meng, L.; Fan, L.; Wu, G.; Ma, L.; Zhao, M.; Huang, Y. Carboxyl functionalization of carbon fibers via aryl diazonium reaction in molten urea to enhance interfacial shear strength. *Appl. Surf. Sci.* **2016**, *362*, 341–347. [CrossRef]

12. Li, N.; Wu, Z.; Huo, L.; Zong, L.; Guo, Y.; Wang, J.; Jian, W. One-step functionalization of carbon fiber using in situ generated aromatic diazonium salts to enhance adhesion with PPBES resins. *RSC Adv.* **2016**, *6*, 70704–70714. [CrossRef]

13. Wang, C.; Chen, L.; Li, J.; Sun, S.; Ma, L.; Wu, G.; Zhao, F.; Jiang, B.; Huang, Y. Enhancing the interfacial strength of carbon fiber reinforced epoxy composites by green grafting of poly(oxypropylene) diamines. *Compos. Part A Appl. Sci. Manuf.* **2017**, *99*, 58–64. [CrossRef]

14. Zho, M.; Meng, L.; Ma, L.; Wu, G.; Xie, F.; Ma, L.; Wang, W.; Jiang, B.; Huang, Y. Stepwise growth of melamine-based dendrimers onto carbon fibers and the effects on interfacial properties of epoxy composites. *Compos. Sci. Technol.* **2017**, *138*, 144–150. [CrossRef]

15. Wu, G.; Ma, L.; Wang, Y.; Liu, L.; Huang, Y. Interfacial properties and thermo-oxidative stability of carbon fiber reinforced methylphenylsilicone resin composites modified with polyhedral oligomeric silsesquioxanes in the interphase. *RSC Adv.* **2016**, *6*, 5032–5039. [CrossRef]

16. Ma, L.; Meng, L.; Wu, G.; Wang, Y.; Zhao, M.; Zhang, C.; Huang, Y. Effects of bonding types of carbon fibers with branched polyethyleneimine on the interfacial microstructure and mechanical properties of carbon fiber/epoxy resin composites. *Compos. Sci. Technol.* **2015**, *117*, 289–297. [CrossRef]

17. Servinis, L.; Henderson, L.C.; Andrighetto, L.M.; Huson, M.G.; Gengenbach, T.R.; Fox, B.L. A novel approach to functionalise pristine unsized carbon fibre using in situ generated diazonium species to enhance interfacial shear strength. *J. Mater. Chem. A* **2015**, *3*, 3360–3371. [CrossRef]

18. Beggs, K.M.; Servinis, L.; Gengenbach, T.R.; Huson, M.G.; Fox, B.L.; Henderson, L.C. A systematic study of carbon fiber surface grafting via in situ diazonium generation for improved interfacial shear strength in epoxy matrix composites. *Compos. Sci. Technol.* **2015**, *118*, 31–38. [CrossRef]

19. Servinis, L.; Gengenbach, T.R.; Huson, M.G.; Henderson, L.C.; Fox, B.L. A novel approach to the functionalisation of pristine carbon fibre using azomethine 1,3-dipolar cycloaddition. *Aust. J. Chem.* **2015**, *2*, 335–344. [CrossRef]

20. Servinis, L.; Beggs, K.M.; Scheffler, C.; Wölfel, E.; Randall, J.D.; Gengenbach, T.R.; Demir, B.; Walsh, T.R.; Doeven, E.H.; Francis, P.S. Electrochemical surface modification of carbon fibers by grafting of amine, carboxylic acid and lipophilic amide groups. *Carbon* **2017**, *118*, 393–403. [CrossRef]

21. Servinis, L.; Beggs, K.M.; Gengenbach, T.R.; Doeven, E.H.; Francis, P.S.; Fox, B.L.; Pringle, J.M.; Pozo-Gonzalo, C.; Walsh, T.R.; Henderson, L.C. Tailoring the fiber-to-matrix interface using click chemistry on carbon fibre surfaces. *J. Mater. Chem. A* **2017**, *5*, 11204–11213. [CrossRef]

22. Eykens, D.J.; Stojcevski, F.; Hendlmeier, A.; Arnold, C.L.; Randall, J.D.; Perus, M.D.; Servinis, L.; Gengenbach, T.R.; Demir, B.; Walsh, T.R. An efficient high-throughput grafting procedure for enhancing carbon fiber-to-matrix interactions in composites. *Chem. Eng. J.* **2018**, *353*, 373–380. [CrossRef]

23. Arnold, C.L.; Beggs, K.M.; Eykens, D.J.; Stojcevski, F.; Servinis, L.; Henderson, L.C. Enhancing interfacial shear strength via surface grafting of carbon fibers using the Kolbe decarboxylation reaction. *Compos. Sci. Technol.* **2018**, *159*, 135–141. [CrossRef]

24. Kümmerer, K. Sustainable chemistry: A future guiding principle. *Angew. Chem. Int. Ed.* **2017**, *56*, 16420–16421. [CrossRef] [PubMed]

25. Raquez, J.; Deléglise, M.; Lacrampe, M.; Krawczak, P. Thermosetting (bio) materials derived from renewable resources: A critical review. *Prog. Polym. Sci.* **2010**, *35*, 487–509. [CrossRef]

26. Dai, J.; Peng, Y.; Teng, N.; Liu, Y.; Liu, C.; Shen, X.; Mahmud, S.; Zhu, J.; Liu, X. High-performing and fire-resistant biobased epoxy resin from renewable sources. *ACS Sustain. Chem. Eng.* **2018**, *6*, 7589–7599. [CrossRef]

27. Li, R.J.; Gutierrez, J.; Chung, Y.-L.; Frank, C.W.; Billington, S.L.; Sattely, E.S. A lignin-epoxy resin derived from biomass as an alternative to formaldehyde-based wood adhesives. *Green Chem.* **2018**, *20*, 1459–1466. [CrossRef]

28. Nazhad, H.Y.; Thakur, V.K. Effect of morphological changes due to increasing carbon nanoparticles content on the quasi-static mechanical response of epoxy resin. *Polymers* **2018**, *10*, 1106. [CrossRef]

29. Thakur, S.; Govender, P.P.; Mamo, M.A.; Tamulevicius, S.; Mishra, Y.K.; Thakur, V.K. Progress in lignin hydrogels and nanocomposites for water purification: Future perspectives. *Vacuum* **2017**, *146*, 342–355. [CrossRef]

30. Laurichesse, S.; Avérous, L. Chemical modifications of lignins: Towards biobased polymers. *Prog. Polym. Sci.* **2014**, *39*, 1266–1290. [CrossRef]

31. Rials, T.G.; Glasser, W.G. Multiphase materials with lignin. VI. Effect of cellulose derivative structure on blend morphology with lignin. *Wood Fiber Sci.* **1989**, *2*, 80–90. [CrossRef]

32. Eykens, D.J.; Servinis, L.; Scheffler, C.; Wölfel, E.; Demir, B.; Walsh, T.R.; Henderson, L.C. Synergistic interfacial effects of ionic liquids as sizing agents and surface modified carbon fibers. *J. Mater. Chem. A* **2018**, *6*, 4504–4514. [CrossRef]

33. Granata, A.; Argyropoulos, D.S. 2-Chloro-4,4,5,5-tetramethyl-1,3,2-dioxaphospholane, a reagent for the accurate determination of the uncondensed and condensed phenolic moieties in lignins. *J. Agric. Food Chem.* **1995**, *43*, 1538–1544. [CrossRef]

34. Kelly, A.; Tyson, A.W. Tensile properties of fiber-reinforced metals: Copper/tungsten and copper/molybdenum. *J. Mech. Phys. Solids* **1965**, *13*, 329–350. [CrossRef]

35. Lopattananon, N.; Kettle, A.P.; Tripathi, D.; Beck, A.J.; Duval, E.; France, R.M.; Short, R.D.; Jones, F.R. Interface molecular engineering of carbon-fiber composites. *Compos. Part A Appl. Sci. Manuf.* **1999**, *30*, 49–57. [CrossRef]

36. Naito, K. Fracture Behaviour of Continuous Carbon Fibre. In *Improvement of Resin Impregnation Property and Reliability Evaluation of CFRP (Carbon Fiber Reinforced Plastic)*; Technical Information Institute Co., Ltd.: Tokyo, Japan, 2010; pp. 34–46.
37. Galli, C. Radical reactions of arenediazonium ions: An easy entry into the chemistry of the aryl radical. *Chem. Rev.* **1988**, *88*, 765–792. [CrossRef]
38. Bahr, J.L.; Tour, J.M. Highly functionalized carbon nanotubes using in situ generated diazonium compounds. *Chem. Mater.* **2001**, *13*, 3823–3824. [CrossRef]
39. Klemm, D.; Heublein, B.; Fink, H.-P.; Bohn, A. Cellulose: Fascinating biopolymer and sustainable raw material. *Angew. Chem. Int. Ed.* **2005**, *44*, 3358–3393. [CrossRef] [PubMed]
40. Ma, L.; Meng, L.; Fan, D.; He, J.; Yu, J.; Qi, M.; Chen, Z.; Huang, Y. Interfacial enhancement of carbon fiber composites by generation 1-3 dendritic hexamethylenetetramine functionalization. *Appl. Surf. Sci.* **2014**, *296*, 61–68. [CrossRef]
41. Zhang, G.; Sun, S.; Yang, D.; Dodelet, J.-P.; Sacher, E. The surface analytical characterization of carbon fibers functionalized by H_2SO_4/HNO_3 treatment. *Carbon* **2008**, *46*, 196–205. [CrossRef]
42. Ehlert, G.J.; Lin, Y.; Sodano, H.A. Carboxyl functionalization of carbon fibers through a grafting reaction that preserves fiber tensile strength. *Carbon* **2011**, *49*, 4246–4255. [CrossRef]
43. Meier, A.R.; Bahureksa, W.A.; Heien, M.L. Elucidating the structure-function relationship of poly(3,4-theylenedioxythiophene) films to advance electrochemical measurements. *J. Phys. Chem. C* **2016**, *120*, 2114–21122. [CrossRef]
44. Marsella, J.A.; Starner, W.E. Acceleration of amine/epoxy reactions with *N*-methyl secondary amines. *J. Polym. Sci. Part A Polym. Chem.* **2000**, *38*, 921–930. [CrossRef]
45. Wu, Q.; Li, M.; Gu, Y.; Wang, S.; Yao, L.; Zhang, Z. Effect of sizing agent on interfacial adhesion of commercial high strength carbon fiber-reinforced resin composites. *Polym. Compos.* **2016**, *37*, 254–261. [CrossRef]
46. Huson, M.G.; Church, J.S.; Kafi, A.A.; Woodhead, A.L.; Khoo, J.; Kiran, M.S.R.N.; Bradby, J.E.; Fox, B.L. Heterogeneity of carbon fibre. *Carbon* **2014**, *68*, 240–249. [CrossRef]

![materials logo] *materials*

MDPI

Article

Seebeck Coefficient of Thermocouples from Nickel-Coated Carbon Fibers: Theory and Experiment

Hardianto Hardianto [1,2,*], Gilbert De Mey [3], Izabela Ciesielska-Wróbel [1], Carla Hertleer [1] and Lieva Van Langenhove [1]

[1] Department of Materials, Textiles and Chemical Engineering, Ghent University, Technologiepark 907, 9052 Zwijnaarde, Belgium; budysiowka@gmail.com (I.C.-W.); carla.hertleer@gmail.com (C.H.); Lieva.VanLangenhove@UGent.be (L.V.L.)
[2] Department of Textile Chemistry, Politeknik STTT Bandung, Jalan Jakarta 31, Bandung 40272, Indonesia
[3] Department of Electronics and Information Systems, Ghent University, Technologiepark 15, 9052 Zwijnaarde, Belgium; Gilbert.DeMey@UGent.be
* Correspondence: hardiant.hardianto@ugent.be

Received: 18 April 2018; Accepted: 25 May 2018; Published: 30 May 2018

Abstract: Thermocouples made of etched and non-etched nickel-coated carbon yarn (NiCCY) were investigated. Theoretic Seebeck coefficients were compared to experimental results from measurements of generated electric voltage by these thermocouples. The etching process for making thermocouples was performed by immersion of NiCCY in the solution containing a mixture of hydrochloric acid (HCl) (37% of concentration), and hydrogen peroxide (H_2O_2) in three different concentrations—3%, 6%, and 10%. Thirty minutes of etching to remove Ni from NiCCY was followed by washing and drying. Next, the ability to generate electrical voltage by the thermocouples (being a junction of the etched and the non-etched NiCCY) was measured in different ranges of temperatures, both a cold junction (291.15–293.15 K) and a hot junction (293.15–325.15 K). A formula predicting the Seebeck coefficient of this thermocouple was elaborated, taking into consideration resistance values of the tested samples. It was proven that there is a good agreement between the theoretical and experimental data, especially for the yarns etched with 6% and 10% peroxide (both were mixed with HCl). The electrical resistance of non-fully etched nickel remaining on the carbon fiber surface (R_1) can have a significant effect on the thermocouples' characteristics.

Keywords: thermocouple; Seebeck coefficient; conductive yarn; nickel-coated carbon fiber

1. Introduction

Smart textiles development has entered a new stage where all the electronic components previously incorporated with textile elements are gradually replaced by electronic-like textiles components, e.g., textile capacitors [1], highly flexible textile antennas [2]. This means that electronic components are usually replaced by pliable and limp films, yarns, fibers, conductive coatings, printed electronics, etc., to make these new smart textiles stretchable, washable, durable and lasting. These last three features have been playing an important role in smart textiles release to the market and the fact that these three conditions have not been fulfilled is exactly why one may not observe all the interesting innovative smart textiles solutions on the market.

This research work is aimed at building a reliable textile thermocouple on a linear textile product (yarn) that could be incorporated as weft, with a thin fabric being a stratum. This stratum that separates two zones provides information on measurements of temperature differences between these two zones. In order to create a textile thermocouple, a junction of two conductive yarns has to be created. In fact, any thermocouple can be created by making a proper circuit with two different conductors that allow the generation of an electric voltage. This electric voltage is known as the Seebeck effect (S).

Its efficiency depends on a ratio of the generated electric voltage and a temperature difference between environments where each of the conductors' join [3].

There are several known solutions attempting textile application of the thermocouple principal presented by other authors, e.g., using constantan with steel to create a thermocouple [4], and constantan with copper for the same purpose [5]. Thermocouples were integrated into textile substrates by researchers as a temperature sensor [6] and heat flux sensor [7–9]. However, there is a disadvantage to the application of these sorts of materials because they are brittle and consequently break easily. Additionally, metal wires integrated into textiles as thermocouples affect the flexibility of these textiles. Therefore, finding textile-based conductive yarns for fabricating a thermocouple and a thermopile for wearable textiles is of great interest.

Recently, huge improvements of the Seebeck coefficient have been found by using other forms of carbon such as carbon nanotube and graphene [10–12]. However, these promising materials are not available in textile yarns which can be inserted into wearable textiles. The toxicity of carbon nanotubes is also a problem. Therefore, nickel-coated carbon yarn (NiCCY) can be a good candidate material for fabricating a textile-based thermoelectric generator.

From our preliminary experiment, a thermocouple from two different conductive textile yarns i.e., carbon fiber and NiCCY demonstrated a Seebeck coefficient of about 18 µV/K. According to this result, there is a possibility to create a thermopile using NiCCY. Carbon fiber coated with nickel is also available on the market. This led us to an idea to create a textile-based thermopile from a single NiCCY, provided that the nickel can be removed selectively to form a series of C-Ni junctions along the NiCCY forming a thermopile to achieve higher voltage output since the voltage generated by a thermopile is proportional to the number of thermocouple junctions. The previous report showed that mixture of peroxide and hydrochloric acid can significantly remove the nickel from nickel-coated carbon fiber [13].

Textile-based conductive yarns are sought for their integration into the textile structures to act as thermocouples or thermopiles, to create more flexible textile-based temperature- or heat flux sensors. There is a great potential for making thermocouples based on existing carbon yarns, especially with Tenax®-J HTS40 A23 12K 1420tex provided by Toho Tenax Europe GmbH, Germany, which has a Ni coating [13]. In this nickel-coated carbon yarn (NiCCY), both C and Ni are electric conductors. Tenax® refers to a group of high-performance carbon fibers. The high tenacity (HT) fibers provide excellent mechanical laminate properties. This fiber is manufactured from a polyacrylonitrile (PAN) precursor and is surface-treated in order to promote adhesion to organic matrix polymers [14].

In this paper, we are going to study the effect of the electrical resistance of the etched NiCCY on the Seebeck coefficient of the thermocouple made of the etched NiCCY and non-etched because it is easier to measure the electrical resistance than to measure the thickness of the nickel layer on NiCCY which is less than 0.25 µm. A formula for predicting the Seebeck coefficient of this thermocouple is going to be elaborated by taking into consideration the resistance values of the yarns. This theoretical formula will be compared to the experimental results.

The electric conductivity σ of the nickel is much higher compared to carbon ($\sigma_{Ni} = 14.3 \times 10^6$ S/m vs. $\sigma_C = 5.9 \times 10^6$ S/m) [15]. In order to make a thermocouple, we have to use two different materials. In this work, this will be performed by etching the Ni layers from the part of the NiCCY, as schematically represented in Figure 1. Both ends ($x = -a$ and $x = b$) are at the "cold" temperature, T_C, which is normally the ambient room temperature. The junction ($x = 0$) between the etched and the non-etched part is at a higher temperature, T_H. At both cold ends, metallic contacts are provided to measure the generated voltage, V_0.

Figure 1 demonstrates an idealised situation in which a single fiber is partially etched, thus having two different amounts of Ni on its surface. The idealised and simplified situation where two different thicknesses of Ni on carbon fiber (CF) are presented, makes the mathematical modeling of a theoretical approach transparent.

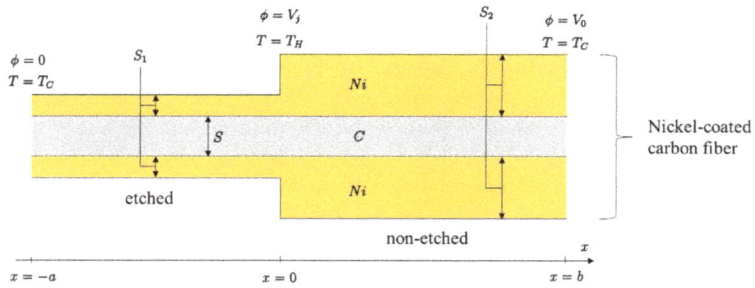

Figure 1. Theoretical diagram of etched and non-etched segment of one filament of nickel-coated carbon fiber as a thermocouple where C is carbon fiber; Ni is nickel coating on the surface of C fiber; S, S_1, and S_2 are the cross-section areas of C, Ni on etched segment and Ni on non-etched segment, respectively; T is temperature; T_C and T_H are cold and hot temperatures, respectively; φ is electric potential; V_j is a voltage at junction (x = 0); V_0 is a voltage at a single end of nickel-coated carbon fiber (x = b); x is axis along the thermocouple.

Theoretically, the most efficient thermocouple is created if a significant part or the whole of the layer of Ni is removed from the surface of CF during the etching process. During the etching process, it was possible to remove an unknown amount of Ni so that when the periodically etched yarn and the non-etched yarn were connected to the voltage meter, the junction allowed for the generation of an output voltage of about 18 µV/K. The etching process and the subsequent measurements were performed on the whole yarn (all the fibers in the yarn were treated at the same time). It would be difficult to treat a single fiber individually. Due to the higher electrical conductivity of Ni, the right part of the fiber section might seem to be made from pure nickel and the presence of carbon may be overlooked. Occasionally, it might happen that not all the nickel has been removed from the part that was intentionally subject to etching so that the performance of the thermocouple might be reduced.

2. Materials and Methods

2.1. Material

Materials used in this experiment were similar to previous work [13]. The NiCCY used in this experiment is Tenax®-J HTS40 A23 12K 1420tex MC from Toho Tenax Europe GmbH, Wuppertal, Germany. Figure 2 shows the yarn and Table 1 presents its parameters.

Figure 2. An image of Tenax®-J HTS40 1420tex NiCCY provided by Toho Tenax Europe GmbH, Wuppertal, Germany. The image was taken with OneBird Smart 5M 300X USB Digital Microscope Camera Video with MicroCapture. The diameter of 6.906 mm was measured without any pretension [14].

Table 1. Characteristics of nickel-coated carbon yarn (NiCCY) [14].

Parameter	Description
Raw material	Carbon
Liner density [tex]	1420 tex
Coating	0.25 μm of Ni
No. of filaments	1200
Filament diameter [μm]	7.5 incl. Coating
Density [g/cm^3]	2.70
Twist [tpm, type]	0
Linear electrical resistance [Ω/m]	2.2667
Commercial name	Tenax®-J HTS40

2.2. Etching Process

The process of removing Ni in this experiment is called an etching process. Combination of HCl and H_2O_2 solution was utilized to remove Ni from NiCCY and is then called etching solution. The H_2O_2 concentration varied from 3%, to 6% up, and to 10%, while HCl was kept constant at 37%. The ratio between H_2O_2 and HCl was 1:1. The samples were immersed in the etching solutions for 30 min. Next, the sample was washed with water to remove the remaining chemicals from the yarn. The excess of water remaining on the sample was absorbed by blotting paper several times and all the samples were air dried at room temperature for a minimum of 24 h prior to testing [13].

2.3. Seebeck Coefficient Measurement

The electric voltage of the thermocouple samples was measured using a nanovoltmeter Amplificateur NV 724 from Setaram, Lyon, France. A Fluke 52 digital thermometer (Fluke, Everett, Washington, DC, USA) was used to measure the temperature during voltage measurement. The higher temperature range for the hot junction was controlled by an electric hot plate, while the lower temperature range for the cold junction was dependent on the external conditions of the laboratory, which were controlled and stable. Figure 3 shows the voltage measurement set up of etched and non-etched NiCCY.

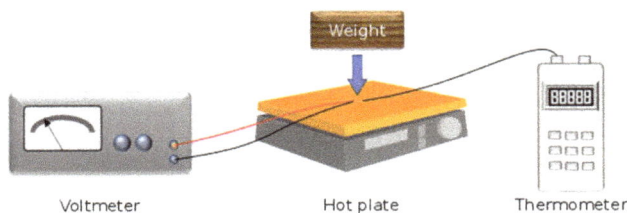

Voltmeter   Hot plate   Thermometer

Figure 3. Illustration of the voltage measurement set up. The junction was placed on the hot plate that had been covered with a piece of paper and a wood weight was placed on the junction and thermometer probe.

The efficiency of the thermocouple was measured through different ranges of temperatures for a cold junction (291.15–293.15 K) and for a hot junction (293.15–325.15 K). These temperature ranges were the only ones enabling the performance of stable and repeatable measurements.

2.4. Theoretical Analysis

The following theoretical calculations are formulated according to the schematic diagram of NiCCY that has been shown in Figure 1 earlier. Generally, the temperature varies along the thermocouple wire: $T(x)$. We know that $T(-a) = T(b) = T_C$ and $T(0) = T_H$. Similarly, the electric potential $\phi(x)$ depends on x. The generated voltage can be calculated as:

$$V_0 = \phi(b) - \phi(-a) \tag{1}$$

Inside each conductor, the current density J (expressed in A/m^2) is given from literature [16]:

$$J = -\sigma\left(\frac{d\phi}{dx} + \varepsilon\frac{dT}{dx}\right) \tag{2}$$

where σ is the electric conductivity and ε the Seebeck coefficient. From the literature, the numerical values are $\varepsilon_{Ni} = -14.8 \ \mu V/K$ and $\varepsilon_C = +3.0 \ \mu V/K$ [17]. Hence, the value $14.8 + 3 = 17.8 \ \mu V/K$ is the highest value one can obtain with a carbon-nickel thermocouple. The electric current I_1 flowing through the left part is then:

$$I_1 = -\sigma_C\left(\frac{d\phi}{dx} + \varepsilon_C\frac{dT}{dx}\right)S - \sigma_{Ni}\left(\frac{d\phi}{dx} + \varepsilon_{Ni}\frac{dT}{dx}\right)S_1 \tag{3}$$

Similarly, for the right part, we have to replace S_1 by S_2 to get I_2:

$$I_2 = -\sigma_C\left(\frac{d\phi}{dx} + \varepsilon_C\frac{dT}{dx}\right)S - \sigma_{Ni}\left(\frac{d\phi}{dx} + \varepsilon_{Ni}\frac{dT}{dx}\right)S_2 \tag{4}$$

The thermocouple voltage is measured with a meter having a high input impedance. Hence, almost no current can flow or:

$$I_1 = I_2 = 0 \tag{5}$$

Integrating (3) with respect to x gives then:

$$\sigma_C(\phi + \varepsilon_C T)S + \sigma_{Ni}(\phi + \varepsilon_{Ni}T)S_1 = A \tag{6}$$

where A is an integration constant. Applying (6) in the point $x = -a$ and $x = 0$ gives, after subtraction:

$$\sigma_C(V_j + \varepsilon_C T_H)S + \sigma_{Ni}(V_j + \varepsilon_{Ni}T_H)S_1 - \sigma_C\varepsilon_C T_C S - \sigma_{Ni}\varepsilon_{Ni}T_C S_1 = 0 \tag{7}$$

or:

$$(\sigma_C S + \sigma_{Ni}S_1)V_j + (\sigma_C\varepsilon_C S + \sigma_{Ni}\varepsilon_{Ni}S_1)(T_H - T_C) = 0 \tag{8}$$

A similar calculation for the right part $(0 < x < b)$ gives us:

$$(\sigma_C S + \sigma_{Ni}S_2)(V_j - V_0) + (\sigma_C\varepsilon_C S + \sigma_{Ni}\varepsilon_{Ni}S_2)(T_H - T_C) = 0 \tag{9}$$

Elimination of the junction potential V_j from (8) and (9) gives us finally:

$$V_0 = (T_H - T_C)\left[\frac{\sigma_C\varepsilon_C S + \sigma_{Ni}\varepsilon_{Ni}S_2}{\sigma_C S + \sigma_{Ni}S_2} - \frac{\sigma_C\varepsilon_C S + \sigma_{Ni}\varepsilon_{Ni}S_1}{\sigma_C S + \sigma_{Ni}S_1}\right] \tag{10}$$

Rewriting (10) gives:

$$V_0 = (T_H - T_C)\frac{\sigma_C\sigma_{Ni}S(S_2 - S_1)}{(\sigma_C S + \sigma_{Ni}S_2)(\sigma_C S + \sigma_{Ni}S_1)}(\varepsilon_{Ni} - \varepsilon_C) \tag{11}$$

As expected, we obtain an output voltage V_0 proportional to the temperature difference $T_H - T_C$. Obviously, the result of Equation (11) is also proportional to the difference of the two Seebeck coefficients, $\varepsilon_{Ni} - \varepsilon_C$.

If $S_1 = 0$, or Ni has been completely removed from the left part, and $\sigma_{Ni} S_2 \gg \sigma_C S$, the relation (11) can be simplified to:

$$V_0 = (T_H - T_C)(\varepsilon_{Ni} - \varepsilon_C) \tag{12}$$

which is the formula we are most familiar with. Note that the term (12) is also the highest value one can obtain with a C-Ni thermocouple.

3. Results and Discussion

3.1. Microscopic Observation after Etching Process

After the chemical treatment of the samples of NiCCY, one observed the effect of the etching process excreted on the treated samples through scanning electron microscope (SEM) image. An untreated sample of NiCCY in Figure 4a was compared to the treated samples in Figure 4b–d where the treatment was 37% HCl and 3%, 6% and 10% of H_2O_2, respectively.

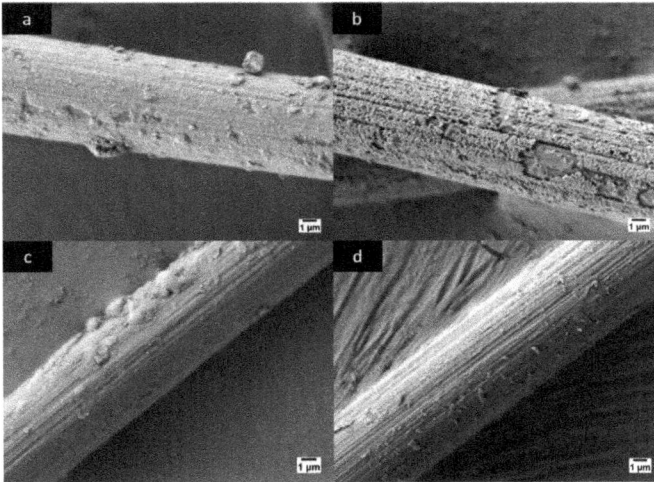

Figure 4. Images from scanning electron microscope (Jeol JSM-7600F) taken with 5000x magnification: (**a**) Untreated nickel-coated carbon fiber; (**b**) nickel-coated carbon fiber after treatment of 3% H_2O_2 + 37% HCl; (**c**) nickel-coated carbon fiber after treatment of 6 % H_2O_2 + 37% HCl; (**d**) nickel-coated carbon fiber after treatment of 10% H_2O_2 + 37% HCl.

Based on the application of different chemical concentration treatments of H_2O_2 to the NiCCY and the microstructure (topography) of untreated and treated NiCCY observed via SEM as shown in Figure 4a–d, one may draw a conclusion concerning the potential impact of the treatment. Namely, the higher the concentration of H_2O_2, the more intense the etching process was, with less Ni remaining on the surface of C. An obvious difference may be noticed between untreated NiCCY (Figure 4a) and the treated one (Figure 4b), where the H_2O_2 concentration was only 3%. In the case of this treatment, the surface of Ni was visibly incised, generating a porous-like layer on the surface of CF. Additionally, some zones of notches on the surface of NiCCY are also visible. This image confirms that in the initial phase, the etching is not homogeneous along the entire yarns, which will also influence the electrical resistance and thermocouple characteristics.

In Figure 4c, one may observe a further etch on the Ni layer due to an increased concentration of H_2O_2 from 3% up to 6%. Although the concentration of H_2O_2 is higher in the case of the treatment presented in Figure 4c, there are still some larger islands of Ni clearly visible. Nevertheless, a large zone of clean and grooved CF is also visible. In the case of the highest utilized concentration of H_2O_2 in this experiment, only some small dust, sparsely distributed, are present on the surface of CF (Figure 4d). Therefore, this confirms that the sample treated in 10% H_2O_2 + 37% HCl can be considered as a fully etched yarn.

3.2. Comparison between Experiment and Theory

In order to check the theoretical analysis, it is more convenient to use resistance values of the yarns to be inserted in (11). The reason is obvious; resistance can be easily measured whereas a cross section like S, S_1 or S_2 are hard to obtain.

The following resistances (expressed in Ω/m) are defined:

$$R_C = \frac{1}{\sigma_C S} \tag{13}$$

$$R_1 = \frac{1}{\sigma_{Ni} S_1} \tag{14}$$

$$R_2 = \frac{1}{\sigma_{Ni} S_2} \tag{15}$$

Equation (11) is then rewritten as:

$$V_0 = (T_H - T_C) \frac{R_C(R_1 - R_2)}{(R_C + R_1)(R_C + R_2)} (\varepsilon_{Ni} - \varepsilon_C) \tag{16}$$

In order to verify the theoretical formula (16), one has to measure the resistance of the yarns involved in our experiment. The results are presented in Table 2 [13].

Table 2. Linear electrical resistance of etched and non-etched yarns [13].

Etching Condition	Linear Electrical Resistance [Ω/m]
Non-etched	2.2667
3% H_2O_2 + 37% HCl (1:1)	2.8667
6% H_2O_2 + 37% HCl (1:1)	31.533
10% H_2O_2 + 37% HCl (1:1)	45.933

From the SEM image, it can be considered that Ni was completely removed in the etching solution containing 10% H_2O_2 + 37% HCl (1:1). The electric resistance is then just the resistance of the carbon or:

$$R_C = 45.933 \; \Omega/m \tag{17}$$

The non-etched part has a resistance of 2.267 Ω/m, which is the parallel connection of R_C and R_2, or:

$$\frac{1}{R_C} + \frac{1}{R_2} = \frac{1}{2.2667} \text{ or } R_2 = 2.3844 \; \Omega/m \tag{18}$$

Similarly, one can calculate the value of R_1 for 6% H_2O_2 and 37% HCl (1:1) etching:

$$\frac{1}{R_C} + \frac{1}{R_1} = \frac{1}{31.533} \text{ or } R_1(6\%) = 100.584 \; \Omega/m \tag{19}$$

and for 3% H_2O_2 and 37% HCl (1:1) etching:

$$\frac{1}{R_C} + \frac{1}{R_1} = \frac{1}{2.867} \text{ or } R_1(3\%) = 3.0575 \; \Omega/m \tag{20}$$

Inserting all the known values into the Equations (12) and (16), one can plot the theoretical graphs as shown in Figure 5 (with lines). The graph for the sample treated with 10% H_2O_2 + 37% HCl was calculated with Equation (12) because S_1 was considered equal to 0. The graphs for the 3% H_2O_2 + 37% HCl and 6% H_2O_2 + 37% HCl were calculated with Equation (16). In Figure 5, the graph of the experimental data (with markers) was also presented as a comparison to the theoretical one.

Values inside the boxes are linear trend line equations for the corresponding data of the 3–10% H_2O_2 with which each was mixed with 37% HCl in 1:1 ratio. The orange and blue boxes are attributed to the experimental and theoretical trend lines, respectively. The Seebeck coefficient values were taken from the slope of voltage vs. temperature difference.

It is clear from Figure 5 that the agreement between the theoretical and experimental data is quite good for the yarns etched with 6% and 10% H_2O_2. The case with 3% H_2O_2 shows a poor agreement. It must be pointed out, however, that 3% H_2O_2 was also a very poor etching. This is proved by the resistance values presented in Table 2. The linear electrical resistance only changed from 2.2667 to 2.8667 Ω/m or 26%.

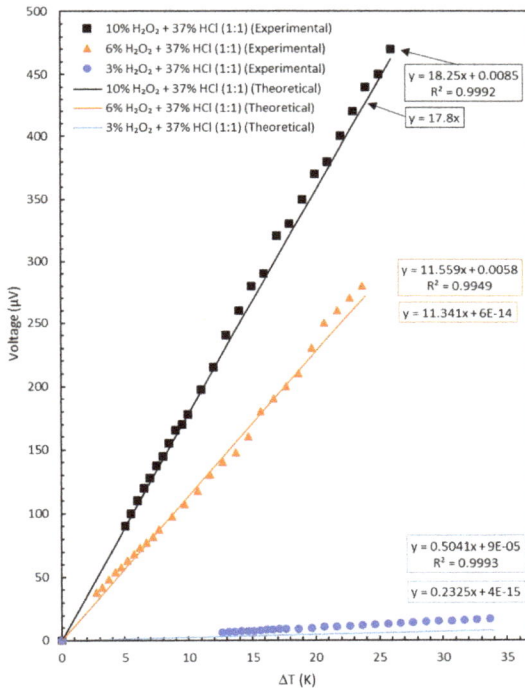

Figure 5. Plot of voltage vs. temperature difference of the samples (theoretical and experimental).

The Equation (16) can be rearranged as:

$$V_0 = (T_H - T_C)(\varepsilon_{Ni} - \varepsilon_C) \frac{\frac{R_C}{R_2}\left(\frac{R_1}{R_2} - 1\right)}{\left(\frac{R_C}{R_2} + \frac{R_1}{R_2}\right)\left(\frac{R_C}{R_2} + 1\right)} \tag{21}$$

If we insert the values from Equations (17)–(20) into the theoretical formula (21), we can make a plot of the generated voltage V_0 versus R_1/R_2 as shown in Figure 6. One observes that the value of R_1 can have a dramatic influence on the overall performance of this thermocouple.

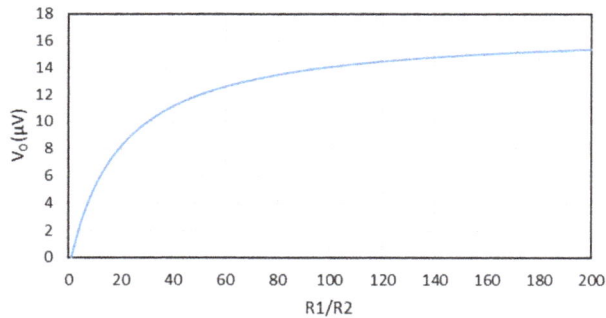

Figure 6. The generated voltage V_0 versus R_1/R_2 plotted from Equation (21) with a temperature difference of 1 K.

4. Conclusions

In this work, a combination of etched and non-etched nickel-coated carbon yarn (NiCCY) was used as the conductive materials for creating textile-based thermocouple. After performing the stripping process in three different concentration of stripping solutions, it is obvious that the higher the concentration of H_2O_2, the more intense the etching process was, with less Ni remaining on the surface of CF. From the theoretical calculation and experimental data, it is proved that there is a good agreement between the theoretical and experimental data, especially for the yarns etched with 6% and 10% H_2O_2 (both are mixed with 37% HCl). The yarn etched with 3% H_2O_2 shows a poor agreement due to the very poor etching action of the chemicals at 3% H_2O_2 and 37% HCl. The 10% H_2O_2 + 37% HCl was efficient enough to etch the Ni from NiCCY. We can conclude that the more efficient the etching process, the better the Seebeck coefficient created from the etched and non-etched NiCCY is. The R-squares of the experimental graphs are all more than 0.99, showing that the data are very close to the fitted regression line. The value of R_1 has a great influence on the whole characteristics of this thermocouple. Overall, we are successful in developing a theoretical formula to calculate the Seebeck coefficient of thermocouples made of etched and non-etched NiCCY based on the resistance value of the samples due to the simplicity of electrical resistance measurement.

Author Contributions: H.H., G.D.M., C.H. and L.V.L. conceived and designed the experiments; H.H. performed the experiments; H.H., G.D.M. and I.C.-W. analyzed the data; H.H., G.D.M. and I.C.-W. wrote the paper; H.H., G.D.M., I.C.-W. and L.V.L. revised the paper.

Funding: This research was funded by the Government of Indonesia through the Indonesian Endowment Fund for Education (LPDP).

Acknowledgments: Hardianto on leave from Politeknik STTT Bandung wants to thank LPDP for the financial support for his stay at Ghent University as a PhD student.

Conflicts of Interest: The authors declare no conflict of interest.

References

1. Choi, C.; Kim, K.M.; Kim, K.J.; Lepró, X.; Spinks, G.M.; Baughman, R.H.; Kim, S.J. Improvement of system capacitance via weavable superelastic biscrolled yarn supercapacitors. *Nat. Commun.* **2016**, *7*, 13811. [CrossRef] [PubMed]

2. Huang, X.; Leng, T.; Zhu, M.; Zhang, X.; Chen, J.; Chang, K.; Aqeeli, M.; Geim, A.K.; Novoselov, K.S.; Hu, Z. Highly Flexible and Conductive printed graphene for wireless wearable communications applications. *Sci. Rep.* **2015**, *5*, 18298. [CrossRef] [PubMed]

3. Uchida, K.; Takahashi, S.; Harii, K.; Ieda, J.; Koshibae, W.; Ando, K.; Maekawa, S.; Saitoh, E. Observation of the spin Seebeck effect. *Nature* **2008**, *455*, 778–781. [CrossRef] [PubMed]

4. Ziegler, S.; Frydrysiak, M. Initial research into the structure and working conditions of textile thermocouples. *Fibres Text. East. Eur.* **2009**, *77*, 84–88.

5. Gidik, H.; Bedek, G.; Dupont, D. Developing thermophysical sensors with textile auxiliary wall. In *Smart Textiles and Their Applications*; Woodhead Publishing: Cambridge, MA, USA, 2016; pp. 423–453.

6. Alexander, R. Jones The Application of Temperature Sensors Into Fabric Substrates. Master's Thesis, Kansas State University, Manhattan, KS, USA, 2011.

7. Dupont, D.; Godts, P.; Leclercq, D. Design of textile heat flowmeter combining evaporation phenomena. *Text. Res. J.* **2006**, *76*, 772–776. [CrossRef]

8. Gidik, H.; Bedek, G.; Dupont, D.; Codau, C. Impact of the textile substrate on the heat transfer of a textile heat flux sensor. *Sens. Actuators A Phys.* **2015**, *230*, 25–32. [CrossRef]

9. Hertleer, C.; Bedek, G.; Dupont, D. A Textile-Based Heat Flux Sensor. In Proceedings of the VI International Technical Textiles Congress, Izmir, Turkey, 14–16 October 2015.

10. Meng, C.; Liu, C.; Fan, S. A promising approach to enhanced thermoelectric properties using carbon nanotube networks. *Adv. Mater.* **2010**, *22*, 535–539. [CrossRef] [PubMed]

11. Esfarjani, K.; Zebarjadi, M.; Kawazoe, Y. Thermoelectric properties of a nanocontact made of two-capped single-wall carbon nanotubes calculated within the tight-binding approximation. *Phys. Rev. B Condens. Matter Mater. Phys.* **2006**, *73*. [CrossRef]

12. Mahmoud, L.; Abdul Samad, Y.; Alhawari, M.; Mohammad, B.; Liao, K.; Ismail, M. Combination of PVA with graphene to improve the seebeck coefficient for thermoelectric generator applications. *J. Electron. Mater.* **2015**, *44*, 420–424. [CrossRef]

13. Hardianto, A.; Hertleer, C.; Mey, G.; De Langenhove, L. Van Removing nickel from nickel-coated carbon fibers. *IOP Conf. Ser. Mater. Sci. Eng.* **2017**, *254*, 72010. [CrossRef]

14. Wróbel, I.C.; Knockaert, J.; Mey, G. De Shielding the electromagnetic waves by inserting conductive lightweight materials into woven curtains. *J. Fash. Technol. Text. Eng.* **2017**, 1–4. [CrossRef]

15. TIBTECH Innovations Properties Table of Stainless Steel, Metals and Other Conductive Materials. Available online: http://www.tibtech.com/conductivity.php (accessed on 1 February 2018).

16. Bejan, A. *Advanced Engineering Thermodynamics*, 3rd ed.; Wiley: Hoboken, NJ, USA, 2006; ISBN 0471677639.

17. McGhee, J.; Korczynski, J.; Henderson, I.; Kulesza, W. *Scientific Metrology*; ACGM Lodart S.A.: Lódz, Poland, 1998; ISBN 83-904299-9-3.

materials

MDPI

Article

Experimental and Numerical Studies on Fiber Deformation and Formability in Thermoforming Process Using a Fast-Cure Carbon Prepreg: Effect of Stacking Sequence and Mold Geometry

Daeryeong Bae [1,2], Shino Kim [3], Wonoh Lee [4], Jin Woo Yi [2], Moon Kwang Um [2] and Dong Gi Seong [5,*]

[1] Advanced Materials Engineering, University of Science and Technology (UST), 217 Gajeong-ro, Yuseong-gu, Daejeon 34113, Korea; drbaeuk@kims.re.kr

[2] Korea Institute of Materials Science (KIMS), 797 Changwon-daero, Changwon, Gyungnam 51508, Korea; yjw0628@kims.re.kr (J.W.Y.); umk1693@kims.re.kr (M.K.U.)

[3] Korea Aerospace Industries (KAI), 78 Gongdanro 1-ro, Sanam-myeon, Sacheon, Gyungnam 52529, Korea; shino.kim@koreaaero.com

[4] School of Mechanical Engineering, Chonnam National University, 77 Yongbong-ro, Buk-gu, Gwangju 61186, Korea; wonohlee@chonnam.ac.kr

[5] Department of Polymer Science and Engineering, Pusan National University, 2 Busandaehak-ro 63beon-gil, Geumjeong-gu, Busan 46241, Korea

* Correspondence: dgseong@pusan.ac.kr; Tel.: +82-51-5102467

Received: 20 April 2018; Accepted: 15 May 2018; Published: 21 May 2018

Abstract: A fast-cure carbon fiber/epoxy prepreg was thermoformed against a replicated automotive roof panel mold (square-cup) to investigate the effect of the stacking sequence of prepreg layers with unidirectional and plane woven fabrics and mold geometry with different drawing angles and depths on the fiber deformation and formability of the prepreg. The optimum forming condition was determined via analysis of the material properties of epoxy resin. The non-linear mechanical properties of prepreg at the deformation modes of inter- and intra-ply shear, tensile and bending were measured to be used as input data for the commercial virtual forming simulation software. The prepreg with a stacking sequence containing the plain-woven carbon prepreg on the outer layer of the laminate was successfully thermoformed against a mold with a depth of 20 mm and a tilting angle of 110°. Experimental results for the shear deformations at each corner of the thermoformed square-cup product were compared with the simulation and a similarity in the overall tendency of the shear angle in the path at each corner was observed. The results are expected to contribute to the optimization of parameters on materials, mold design and processing in the thermoforming mass-production process for manufacturing high quality automotive parts with a square-cup geometry.

Keywords: thermoforming; prepreg; carbon fiber; fast-cure epoxy resin

1. Introduction

As fuel economy regulations in the automotive industry become tighter [1], carbon fiber reinforced plastic (CFRP) parts and their processing are more attractive because the parts have lighter weight with better mechanical properties than steel parts while still maintain the durability and safety as the vehicle parts. To commercialize CFRP parts, a cost-effective mass production line must be developed. The ideal cycle time for a production line would be between 3 and 7 min per part. One of the forming technique that meets the above criteria is prepreg compression molding (PCM), which is also known as a thermoforming (or hot stamping) process [2]. Several studies have investigated the formability of a carbon fiber/thermoplastic prepreg on a hemisphere. The behavior of the fiber reorientations due to the fiber deformation from the two-dimensional prepreg to the three-dimensional mold shape during the thermoforming process [3,4] was studied to observe mechanical properties of the final product. The studies concluded that the fiber deformation behavior in the formed product was determined by the fiber orientation of the prepregs and the mold profile. Thermoforming processes with asymmetric L-shape part and square box shape were also studied to compare the computed shear angle values with the experimental results [5,6]. Analysis of thermomechanical behaviors such as in-plane shear and bending stiffness for thermoplastic based composites was conducted to understand the formation of wrinkles in the final thermoformed product [7–9]. Non-linear thermomechanical properties of prepreg at various deformation modes including frictional stress as intra-ply shear were also investigated [10–12]. Some benchmark studies were also carried for standardization of the test methods for measuring thermomechanical properties of prepreg in order to improve the accuracy of thermoforming process simulation [13,14].

However, only a few studies on fast thermoforming process by using fast cure thermoset resin were identified and there was a lack of research on the effect of the formability of the carbon fiber/epoxy prepregs on variations in the mold geometry and stacking sequence of laminate [15]. In automotive industries, the geometries of the CFRP-based vehicle parts, such as the roof, hood and trunk, are based on a flat quadrangular (square, rectangular or trapezoid) plane with extreme curvatures on the edge. The curvatures are known to be the most challenging section of the part to form and it is likely to create distortions during thermoforming process. Consequently, experimental studies aimed at high-speed compression molding of automotive parts are required to investigate the formability of the various stacking sequences of prepreg with different mold geometric shapes using a fast-cure epoxy resin. A simulation of forming process is also necessary to predict the formability of the part, which leads to the optimization of the process parameters for defect-free thermoformed product in advance.

In this work, highly accurate methods and testing devices to measure the non-linear mechanical behaviors of fast-cured carbon or glass fiber/epoxy prepregs with different woven patterns were prepared, which were used to measure the tensile, intra- and inter-ply shear properties of prepregs. The optimal processing temperature and curing time for the prepreg were determined by measuring the rheological and thermal properties of the resin [16]. Two different mold types with square-cup geometries were prepared and thermoforming tests were conducted to evaluate the fiber deformations and formability of the prepreg based on the draft angle and depth of the mold. After the forming stage, the in-plane shear deformation in the final parts was quantified. The commercial virtual forming simulation software (PAM-FORM) with input data obtained from the mechanical property tests was used to simulate the effect of the square-cup geometry and different stacking sequences of the prepreg layers on the shape and property of the final product, which was also compared to the experimental results. It is expected that our cumulative experimental and simulation results can provide a design guide for the fast-cure type prepreg thermoforming process.

2. Experimental

2.1. Materials and Sample Preparation

Three types of prepreg were provided by Hankuk Carbon Co., Ltd., Miryang, Korea. The specifications of the prepregs are shown in Table 1. The prepregs consisting of plain woven (PW), unidirectional (UD) carbon or glass fibers were specifically formulated with a fast-cure, BPA-type thermosetting epoxy resin.

In the forming stage, ten-layered laminates with symmetrical stacking sequences with respect to the mid-plane were prepared. Each layer was cut into a 300 mm × 300 mm square at various fiber angles (0/90°, −45°/+45°). Fiber angles were measured based on the rolling direction of the prepreg as the reference point. Table 2 illustrates two different laminate stacking sequences for verifying the formability of the square cup. Ten repeated measurements for the thickness of each prepreg were conducted by using an optical microscope (ECLIPSE LV150N, Nikon, Tokyo, Japan).

Table 1. Physical properties of the different types of prepreg.

Model Name	Type	Thickness (Standard Deviation) (mm)	Resin Content (vol %)	Weight (g/m^2)
7628	Plain woven (PW) glass	0.305 ($\pm 3.1 \times 10^{-6}$)	42	209
CF-3327	Plain woven (PW) carbon	0.269 ($\pm 5.7 \times 10^{-6}$)	42	200
CU-190	Unidirectional (UD) carbon	0.202 ($\pm 3.8 \times 10^{-6}$)	38	190

Table 2. Description of the stacking sequences for thermoforming tests.

Ply	Pattern 1	Pattern 2
1	Carbon PW	UD 0°
2	UD 0°	UD +45°
3	UD +45°	UD −45°
4	UD −45°	Glass PW
5	Glass PW	Glass PW
6	Glass PW	Glass PW
7	UD −45°	Glass PW
8	UD +45°	UD −45°
9	UD 0°	UD +45°
10	Carbon PW	UD 0°

2.2. Measurement of the Non-Linear Mechanical Behaviors of the Prepreg at Elevated Temperatures

The optimum temperature for prepreg forming process is the temperature at which the resin has the lowest viscosity. Mechanical behavior of prepregs during the tensile [13], in-plane shear (intraply) [14] and frictional (interply) tests [10] may be affected by the flowability of the resin at elevated temperatures. Therefore, the developed methods and apparatus were introduced to characterize the different types of prepreg.

Although standardized methods to measure the tensile properties of the prepreg at elevated temperatures are unavailable, the specimen with an appropriate size were prepared for the tensile test. The overall length of the specimen and the gauge length were fixed at 240 mm and 100 mm, respectively. All the tests were performed at a cross-head speed of 10 mm/min and a chamber temperature of 100 °C. To minimize the temperature variation of the specimen, chamber was pre-heated for 5 min. After fixing the specimen in the grips, temperature was stabilized until required temperature was set. For the UD 90° carbon specimen, the tests were conducted under a low load using a dynamic mechanical analysis (DMA). The temperature stabilization procedure for the specimen was conducted for 5 min before starting the experiment. To overcome slippage at the end of the specimen during the tensile test, glass reinforced epoxy tabs with sandpaper were attached to both sides of the fabric, which can increase the frictional force between the tab and the fabric.

Several layers of prepreg require that the prepreg deforms against a particular design geometry without any wrinkles. This can be achieved via intraply shear within the prepreg ply and interply shear in-between the prepreg plies [13]. Interply shear deformation due to friction can be described by measuring the coefficient of friction for the prepreg-prepreg and prepreg-mold surfaces. The forming resistance created by a high coefficient of friction can cause wrinkles and built-in residual stresses within a stack of prepreg [14]. A suitable model to describe the interply shear deformation via friction at an elevated temperature is the hydrodynamic model [15]. The coefficient of friction is required to measure the friction between the two plies separated by a thin layer of low viscosity resin (fluid). Frictional force between the specimen and the tool surface was measured by using the as-made measuring apparatus with load cell, a pneumatic cylinder and step motor, which has been described in our previous research [17]. Preheating procedures for specimen was required to provide a constant viscosity of resin inside the prepreg. The mold temperature was set at 100 °C for a constant epoxy resin viscosity. The specimens were prepared with a length of 180 mm and a width of 85 mm and attached to both sides of the steel frame. To define the fiber orientation angle of the specimen, the horizontal movement direction of the test frame was designated as a reference line.

A bias extension test [7,8] was conducted to identify the intra-ply shear properties of three different prepregs under the forming condition of 10 mm/min at 100 °C. Cross-piled fabrics +45° with −45° of UD, PW carbon and PW glass (50 mm × 240 mm) were used to keep the specimen in a symmetrical shape during the tensile test using the Shimadzu test machine and heating chamber (Figure 1a,b). Procedures for preheating and temperature stabilization inside the chamber were identical with the high temperature tensile tests.

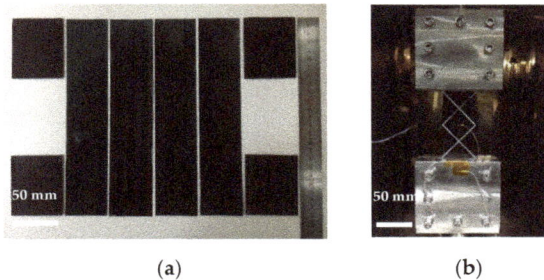

(a)   (b)

Figure 1. (a) Sample preparation and (b) test apparatus for the bias extension test.

After obtaining shear angle from bias extension test, force-shear angle curve can be plotted. In order to represent the relationship between the normalized shear force and shear angle, the torque per original unit area can be calculated by using Equation (1) and an iterative process with Equation (2) [14]. The shear modulus can be calculated by measuring the slope of the shear stress-strain curve.

$$C_s(\gamma) = F_{sh}(\gamma) \cdot \cos \gamma, \tag{1}$$

$$C_s(\gamma) = \frac{1}{(2H - 3W)} + \left(\left(\frac{H}{W} - 1 \right) F \left(\cos \frac{\gamma}{2} - \sin \frac{\gamma}{2} \right) - WC_s \left(\frac{\gamma}{2} \right) \right), \tag{2}$$

where $C_s(\gamma)$ is the torque per original unit area that is needed to deform the fabric in shear, F_{sh} is the normalized shear force and F is the power made through the clamping force.

The bending stiffness of the carbon fiber reinforced thin prepreg is generally very low compared to tensile strength due to the sliding between fibers and between yarns [18]. Therefore, the bending resistance of prepreg is considered as a less important factor during the forming process. Membrane element approaches were also used in macroscopic forming simulation to neglect the bending stiffness [19]. However, Liang et al. suggested that bending stiffness was a significant factor to describe wrinkle formation during the forming simulation. The author also proved that the number and size of the wrinkles obtained in the simulation results were caused by different level of the bending stiffness [20]. In this article, the formability of the prepreg with different tilting angles and material lay-ups has been emphasized to analyze especially corner and edge sections of thermoformed product rather than the formation of wrinkles on the main outside section of the part. Since the magnitude of bending stiffness is required as input data for simulation, results of calculated bending stiffness by a vertical cantilever test with a linear actuator and load cell at 90 °C from Alshahrani et al. for woven carbon/epoxy prepreg were used in the forming simulation [21]. Rheological behavior of epoxy resin and test parameters such as temperature and forming speed used in Alshahrani's work were very similar to this work, which enabled us to use their test results on bending stiffness.

2.3. Mold Geometry

Two types of molds with a rectangular shape were prepared with tilting angles and drawing depths to investigate the effect of the geometric shape of the mold and punch on the prepreg formability. It was thought that the formability and demoldability of the prepreg would be affected by the drawing depth and angle. The drawing depth and tilting angle for the square-based Type 1 mold were 20 mm and 110°, respectively. The maximum width of the Type 1 mold was 100 mm and the minimum width at the bottom of the mold was 85.44 mm. A 100 mm × 100 mm square-based Type 2 mold had a 40 mm drawing depth with a 90° tilting angle. The corner radii of both the mold and square punch were 4.7 mm and 2.7 mm, respectively. The clearance between the punch and the mold was 2 mm, which was the target thickness for the thermoformed square-cup.

2.4. Thermoforming Apparatus and Procedure

The overall experimental apparatus for the thermoforming of the mold is illustrated in Figure 2. A 250 kN universal testing machine with an environmental convection chamber made by Instron 5985 (Instron, Norwood, MA, USA) was used to provide a constant punch forming speed under controlled environmental conditions. Temperatures of the electrically heated blank holder, mold, punch and chamber were individually controlled using cartridge heaters, which were kept at 100 °C. It was thought that thermoforming process with an isothermal condition would be advantageous to high-speed manufacturing process for automotive parts. If thermoforming process with multi-stage temperature conditions were used in the automotive industries, time required to raise and lower the mold temperature would be much longer than isothermal conditions. The forming speed was set to 10 mm/min and a constant weight of 10 kg (blank holder) was used throughout the experiment to minimize the wrinkles on the surface of the prepreg. The lubricant demolding agent for the thermosetting polymer was applied to the mold, punch and blank holder were dried at 60 °C for 30 min.

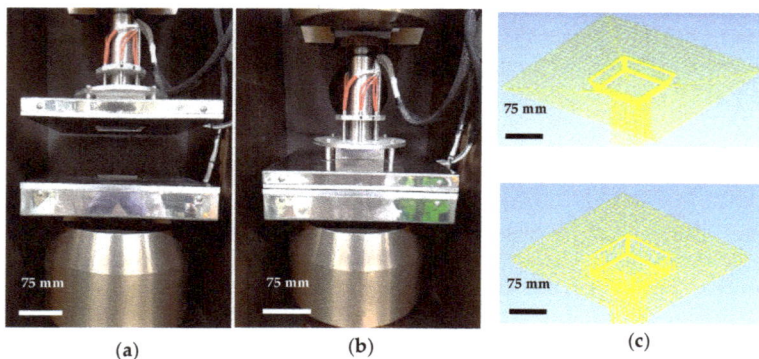

(a) (b) (c)

Figure 2. Thermoforming experimental apparatus for (**a**) open, (**b**) closed square-cup mold, (**c**) Type 1 with a 20 mm mold thickness and 110° draft angle (**top**) Type 2 with a 40 mm mold thickness and 90° draft angle (**bottom**).

The procedure of prepreg square cup isothermal forming test is as follows. Ten-layered laminates of prepreg with the previously described stacking sequence were placed on the lubricant-coated base mold. The blank holder and punch were positioned at 2 mm from the top of the prepreg layer surface. The displacement of the punch was set to zero and the temperature of the mold, blank holder, punch and chamber was raised to 100 °C. As soon as the temperature reached 100 °C, the forming started with a 10 mm/min punching speed until the desired depth of the mold was reached. The final stage was to cure the prepreg at 120 °C for 10 min and cool to room temperature for demolding.

2.5. Forming Simulation

In order to fabricate a complex shape of final product without defects, the process simulation should be conducted. One of the challenges in thermoforming is the fiber deformations during the process, which may affect the mechanical performance of the final composite part. The commercial finite element (FE) code, PAM-FORM, provides a validated methodology and constitutive model for describing unidirectional and fabric sheets pre-impregnated by thermoset or thermoplastic resin. The composite material model in the PAM-FORM was used to describe the realistic behavior of prepreg including the fabric and unidirectional fiber structure [22–24]. Pre and post locking shear modulus can be defined by locking angle between warp and weft direction of fiber due to the shear force. Viscous friction law may be considered with temperature variations. Maxwell model in parallel with two linear elastic fiber phases is used for the mathematical material model. Heat transfer within the prepreg ply and ply-tool and heat conduction through shell elements are also modeled [25]. The constitutive law for plane stress of textile and resin materials can be defined as Equation (3).

$$\begin{pmatrix} \sigma_{11} \\ \sigma_{22} \\ \sigma_{12} \end{pmatrix} = \begin{pmatrix} E_{11} & \varepsilon_{11} \\ E_{22} & \varepsilon_{22} \\ G_{12} & \varepsilon_{12} \end{pmatrix} + \begin{bmatrix} 4\eta_L & 2\eta_T & 0 \\ 2\eta_T & 4\eta_L & 0 \\ 0 & 0 & 2\eta_L \end{bmatrix} \begin{pmatrix} \dot{\varepsilon}_{11} \\ \dot{\varepsilon}_{22} \\ \dot{\varepsilon}_{12} \end{pmatrix}, \tag{3}$$

The first term of this equation is related to the elastic contribution of fiber and the second term includes the longitudinal (η_L) and transverse (η_T) terms of viscosity in either isothermal or temperature-dependent forming conditions.

3. Results and Discussion

3.1. Material Properties of the Resin

The optimum forming conditions were determined using differential scanning calorimetry (DSC, TA Q2000) and a rheometer (Anton Paar Modular Compact Rheometer 302) to identify the curing temperature time of the matrix in the prepreg. The epoxy resin began to cure after 110 °C in the temperature sweep test of the viscosity and the lowest viscosity region before the curing stage was between 95 and 105 °C, as shown Figure 3a. Therefore, 100 °C was selected as the thermoforming temperature for the epoxy/carbon prepreg because it was the temperature with the lowest viscosity. The DSC dynamic and isothermal scanning results provided the relationship between the degree of cure and time at variable temperatures (Figure 3b). The degree of cure (Equation (4)) is the ratio of the isothermal heat reaction (H_T) to the amount of heat generated during dynamic scanning (H_U). Resin is considered to be uncured when $\alpha = 0$ (0%) and fully cured when $\alpha = 1$ (100%) [26].

$$\alpha = \frac{H_T}{H_U} \tag{4}$$

When the epoxy resin was exposed to a heat of 150 °C, the degree of cure was calculated as 98.5%, while was 46.6% and 64.7% at 110 °C and 130 °C, respectively. It was found that 100 °C had the lowest degree of cure (45.8%) among the 4 different temperature ranges throughout the curing process as shown in Figure 3b. When the polymer chains in of a fast-cure type epoxy resin passed at a specific point (temperature or time), a rapid cross-linking actions between the chains led to the establishment of network structures that provided the limitation on formability of prepreg.

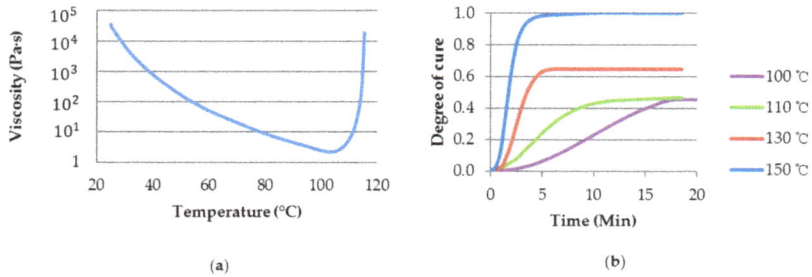

(a)   (b)

Figure 3. (**a**) Viscosity dynamic scan; (**b**) Degree of cure versus time at different temperatures for the fast-cure epoxy resin.

3.2. Mechanical Properties of Prepreg at the Process Temperature

The tensile strength and Young's modulus of three different prepregs were measured to demonstrate the non-linear mechanical behavior of prepreg at high temperature forming conditions. The thickness of each fiber was averaged by measuring 10 different points along the thickness direction. It was observed that UD and PW fabrics were fully impregnated by epoxy resin without any major voids, which might not affect the overall thickness of prepreg in high temperature conditions. Therefore, the assumption was made that thicknesses of three different prepregs at high temperature with the lowest viscosity were considered to be identical to the thickness of prepregs at room temperature. The thickness measurements showed that prepregs containing plain woven pattern fabric were thicker than UD carbon prepregs. Consequently, the thickness of prepreg was highly dependent on the structure of fabric.

Load and displacement for the prepreg was measured using a universal test machine by three times each and the stress-strain curve and the mechanical properties for each prepreg are shown in Figure 4a and Table 3. The tensile stress-strain curves for UD 0°, as illustrated in Figure 4a, show that the rigid epoxy resin within the fiber bundles at room temperature exhibits more linear behavior than high temperature of 100 °C and resultantly improves the overall tensile properties of the material. Non-linear mechanical properties of prepreg are also caused by the weaving pattern of reinforced fiber and these intrinsic behaviors of plain woven carbon and glass prepreg may be strengthened by an additional flow of resin at a high temperature. The slope of intermediate linear region in the stress-strain curve was chosen to calculate the modulus of each prepreg. Although the values of tensile strength and modulus were mentioned in Table 3, they were not able to represent the non-linear behavior of prepreg at an elevated temperature during the thermoforming simulation. Therefore, five data points averaged from the five distinguished regions of each mechanical test for prepregs were used as input data for non-linear mechanical properties in the simulation. No slippages were identified after the test and fiber fractures were observed near the grip of the specimen (Figure 4b). The mechanical properties of UD carbon 90° were measured using dynamic mechanical analysis (DMA) and its characteristic was represented by only resin properties, which were the lowest values recorded among the four prepregs. Therefore, fiber separation in the outer layer of pattern 2 containing the UD carbon is likely to occur during the forming process.

Table 3. Average and standard deviation of strength and Young's modulus from three repeated measurements for unidirectional (UD) 0° (red), PW carbon (green), PW glass (blue) and UD carbon 90° at 100 °C.

Prepreg	UD Carbon 0°	PW Carbon	PW Glass	UD Carbon 90°
Strength (MPa)	280.1 (±18.3)	72.2 (±5.3)	30.0 (±1.1)	1.4×10^{-3} (±1.6×10^{-4})
Modulus (GPa)	60.4 (±3.7)	11.2 (±1.1)	0.57 (±0.04)	2.9×10^{-2} (±2.6×10^{-3})

The COF for three different prepregs were evaluated with the thermoforming parameters. Table 4 shows that a suitable prepreg layer against the tool surface is included in the pattern containing the outer UD carbon layer and this layer would minimize the internal stresses of the laminate during the forming process [27]. However, a high risk of fiber separation for the UD carbon during the forming must be considered. An identical experimental procedure for the prepreg-tool surface was conducted to evaluate the prepreg-prepreg interaction with five different inter-ply combinations (Table 5). Both sides of the steel frame were used to fix the prepreg and the top and bottom surfaces of the electrically heated steel mold were covered with another type of prepreg. Three repeated measurements of friction force for five different fabric patterns were conducted. A similar trend was observed in the prepreg-prepreg patterns containing the PW carbon and glass prepregs with prepreg-tool interactions, which exhibited a higher COF for the PW carbon and glass (No. 1, 4 and 5 in Table 5) than the patterns containing only UD carbon (No. 2 and 3). Different fiber orientations between the UD carbon prepregs also increased the COF compared to the frictional force between the UD carbon and tool surface. This is because of the higher frictional force created by the different direction fibers sliding against each other. Therefore, the frictional behavior of the laminate during the forming stage is mainly dependent on the weave patterns and fiber orientations of the prepreg. Moreover, the adjacent plies containing PW carbon or glass might help preventing any fiber separations of UD prepregs that could lead to a reduction in wrinkles because of their high COF [28].

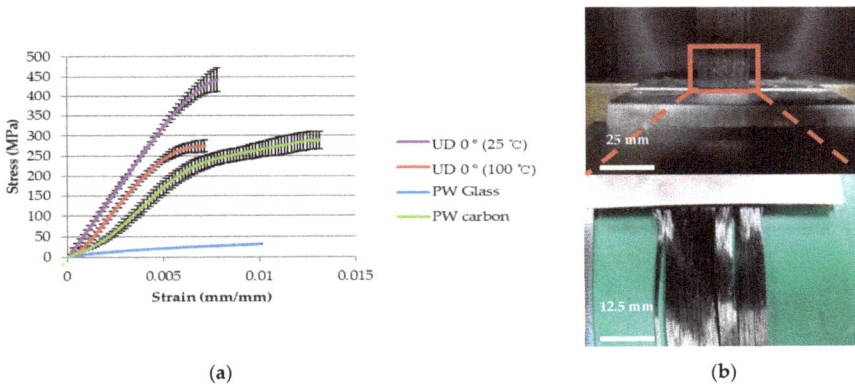

(a)   (b)

Figure 4. (a) Non-linear stress-strain curves for each prepreg at 100 °C; (b) Enlargement of UD 0° prepreg images after a high temperature tensile test.

Table 4. Coefficient of friction between tool and prepreg with three repeated friction force measurements at the thermoforming conditions.

Prepreg	UD Carbon 0°	UD Carbon 90°	PW Carbon
Coefficient of friction	0.050	0.063	0.202
Standard deviation	0.007	0.010	0.011

Table 5. Coefficient of friction of five different prepreg-prepreg patterns against our process parameters.

No.	Interface	Coefficient of Friction (COF)	Standard Deviation
1	PW carbon/UD 0°	0.180	0.002
2	UD 0°/UD +45°	0.078	0.002
3	UD +45°/UD −45°	0.074	0.005
4	PW glass/UD −45°	0.162	0.004
5	PW glass/PW glass	0.157	0.004

Three different prepregs were characterized using a bias extension test to study their in-plane shear behaviors under the forming conditions with a constant speed. All cross-plied specimens were deformed symmetrically and three obvious shear zones within the specimen were identified via the bias extension tests. Optical analysis of shear angle measurements obtained from the image analyzer software (Figure 5) was also conducted to compare with theoretical shear angles calculated from Equation (1). The results showed that the measured values of shear angle from three different prepregs were within the theoretical value ranges as illustrated in Figure 5. The possibility of lubricating effect in the intersections of warp and weft fiber by the resin flow would affect the shear angles obtained from the optical analysis [29].

During the load-displacement measurements, the initial force variation after three repeated measurements was low until the displacement was 6 mm and then the variation increased gradually towards the end of the experiment. The shear stress-shear strain curve for each prepreg as illustrated in Figure 5 shows that the shape depends on the weave patterns of the prepreg. The results reveal that the ideal shape of the curve in the highly shear-dominated PW carbon contains three distinctive regions including a shear locking area with two inflection points along the curve. The first inflection point is located where the shear stress is approximately 0.01 MPa and the shear strain is 0.0747 rad, which indicates a sharp increase in the stiffness during the yarn rotation 1.34×10^{-4} (GPa). The rotation between the warp and weft due to the shear force is limited when shear strain is 0.403 rad, which is also known as the shear locking angle [30]. The shear modulus in the shear locking angle region decreased to 2.01×10^{-5} GPa. Deformation beyond the locking angle leads to wrinkles in the specimen, which is caused by out-of-plane deformation or buckling. After passing second inflection points, a fractional increase in the shear modulus of 4.78×10^{-5} GPa was then observed upon aligning the reoriented yarns in the loading direction. The shear angle range below 0.8 rad was considered in this study, because the scattering range of the obtained shear modulus values above 0.8 rad was significantly wide to use as input data in thermoforming simulation. On the other hand, a coarse and irregular plain woven pattern of glass (PW glass) causes more indefinite inflection points and shear locking areas in the shear stress-shear strain curve in comparison with PW carbon. In the case of the UD carbon, the plateau region of the graph after a sharp increase in the stiffness indicates that the fabric straightens without interference between the bundles [10].

3.3. Square-Cup Drawing Test

The formability of the prepreg laminate was analyzed via drawing tests using the specified design of the base mold, punch and holder. Before the main forming operation, the input for the main temperature controller was set at 100 °C for the punch and chamber, 90 °C for the holder and 85 °C for the base mold to ensure the set temperature was not exceeded.

The views from both sides of the thermoformed square-cup with the pattern 1 prepreg laminate using the Type 1 mold are shown in Figure 6. The dimensions of the prepreg laminate in Figure 6b also show that the middle sections of each side line were drawn into the center by a few millimeters. This contraction made it possible to demold the formed square-cup without any significant external stresses. Flat and smooth surfaces were obtained on the inclined wall of the square-cup after demolding. However, excessively cured epoxy layers at the corners and edges were present on the outer and inner layer, respectively. The large resin content on the surface of the laminate (PW carbon) and the high flow of the resin-rich areas inside the prepreg laminate during the thermoforming caused the highly pressed epoxy resin to agglomerate at the corner of the square-cup shaped part. On the other hand, an identical prepreg pattern laminate with the Type 2 mold design shown in Figure 7 involved a deeper drawing action. More wrinkles and shrinkages around the square-cup were observed. The process was found to be more time-consuming than type 1 mold for demolding the final formed product. It was thought that shrinkage of formed product was likely to occur around the horizontal centerline of product. This would create an increased amount of contact pressure and friction between the product and mold if a proper tilting angle was not provided. Therefore, determination of the ideal tilting angle for a given mold was very important factor to improve the demolding process. The square-cup with the Type 2 mold had more excessive epoxy resin at the edges of the inner and outer layers than the Type 1 mold. Therefore, to meet the Class A surface quality for automotive products, a non-uniform resin distribution on the surface of the formed product due to the resin-rich area must be taken into consideration when the compositions of the prepregs and design of molds are determined. Although there were some difficulties to identify the shear deformation in the formed pattern 1 against Type 2 mold, Wang et al. [31] studied the yarn reorientation of plain woven carbon fabric with square-shaped punch during the preforming process. The study showed that shear angles at the corner of square box were between 50 ° and 60 °. Therefore, high shear angles in pattern 1 with Type 2 mold would had been affected by a deeper drawing action. When pattern 2, which contained UD carbon prepreg on the outer layer, was formed against a Type 1 mold, fiber separations and distortions (wrinkles) occurred on both sides of the square-cup (Figure 8). This is due to the orientation of the UD carbon at 90 ° with respect to the punching direction (downwards). Only shrinkage of the fiber orientation at 0° (left and right side of the laminate in Figure 8b) appeared after thermoforming. The amount of the excessively cured resin layers was reduced at the corners and edges of the square-cup. Therefore, the amount of resin on the outer layer of the pattern 2 laminate can be regarded as the optimum amount of resin in the thermoforming process to prevent epoxy resin being agglomerated. However, from an aesthetic point of view, pattern 2 is not the optimum composition to thermoform the square-cup.

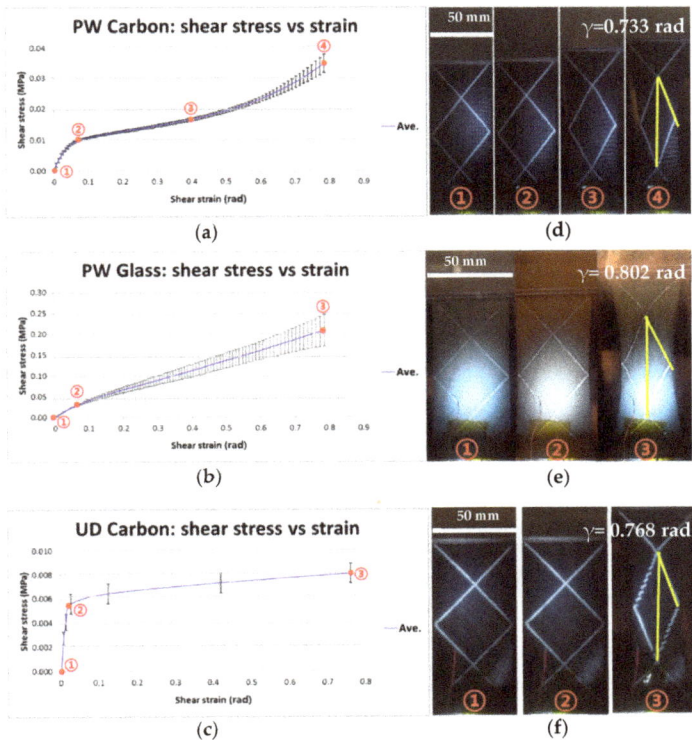

Figure 5. Comparison between the theoretical shear strain (angle) of for (**a**) Plain–woven (PW) carbon; (**b**) PW glass and (**c**) UD carbon prepreg and real shear angle measurements; (**d**) PW carbon; (**e**) PW glass and (**f**) UD carbon prepreg at a high temperature.

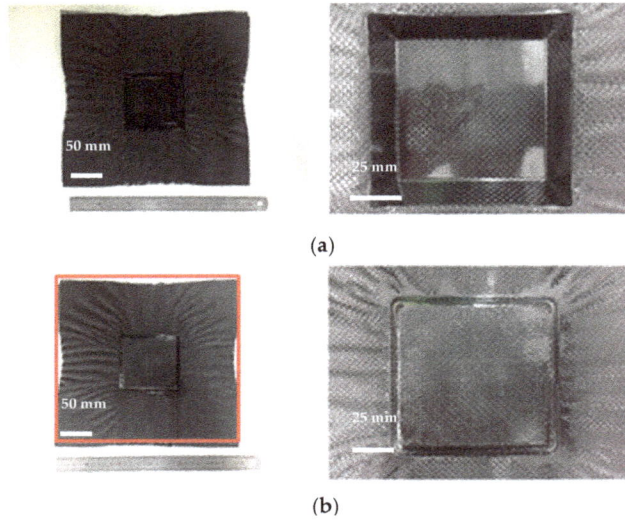

Figure 6. (**a**) Inner layer and (**b**) outer layer of the thermoformed square-cup for the Type 1 mold design with the pattern 1 prepreg laminate.

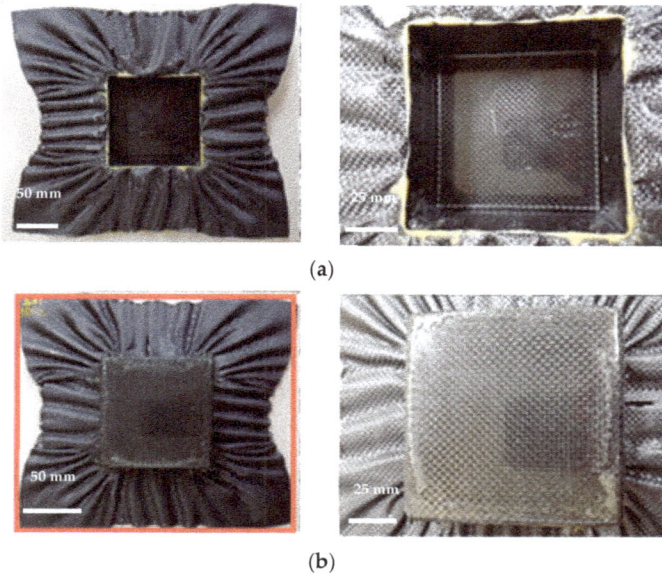

Figure 7. (**a**) Inner layer and (**b**) outer layer of the thermoformed square-cup for the Type 2 mold design with the pattern 1 prepreg laminate.

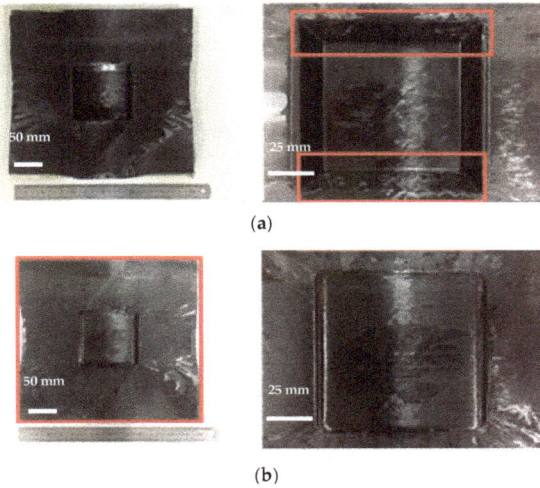

Figure 8. (a) Inner layer and (b) outer layer of the thermoformed square-cup for the Type 1 mold design with the pattern 2 prepreg laminate.

The evidence at the four corners of the square-cup suggested that the formability of the outer layer of the PW carbon prepreg in the pattern 1 laminate was highly shear-dominated during thermoforming. Three different regions at the corner of the outer PW carbon layer are illustrated in Figure 9 and the deformations of the yarns in the PW carbon gradually shifted towards the center line of the corner where the most severe deformation occurred via shear action (Region ③ in Figure 9). However, the variations in the upper and lower angles of the deformed fabric yarns in the inner and outer layer were too wide to use as conclusive results. This might be caused by the asymmetric shear deformations of the fabric during the thermoforming process. It was thought that the scattered data also depended on the uniformity of the weaving strength and number of yarns per bundle in the PW carbon. The average values of the upper and lower angle in each rhombic shape were used to calculate the shear angles. Some of the yarns in the inner layer slipped out of place and the discontinuity of the rhombic patterns was identified in the region where the most extreme shear was experienced (Figure 9a). To estimate the angle in this case, the mid-point between two rhombic shapes was designated by connecting imaginary lines in a diagonal direction to measure the angle between the warp and weft yarns (Figure 9b). A general trend of the experimental results shows that the shear angle increases for the first 10 mm of the path and reaches the highest value between 10 mm and 15 mm on the path. The shear angles between the outer and inner layers of the PW carbon did not coincide each other at each corner of the square-cup.

The virtual forming simulation for thermoforming of the square-cup was implemented in PAM-FORM to compare the deformation behavior of the prepreg laminate with the experimental data. The most severe shear deformations experienced during the forming simulation were represented by the red and blue zones as illustrated in Figure 10 and the predicted shear angle in each element was considered as an absolute value. Each element (rhombic shape) was then analyzed to find the major diagonal distance (path length) and shear angle. The refinement level of mesh at the center of material layers where meets punch was higher than outside of the square cup in order to achieve high accuracy of shear deformations at the corners of square cup. This could be the one of the reason why wrinkles were not predicted in the outside region, as illustrated in Figure 10. An attempt to simulate the Type 2 mold was unsuccessful due to the high number of wrinkles around the square-cup from the deeper drawing action at the center of the laminate, which was seen in the experimental results.

The simulation data had a tendency to increase the shear angle along the corner of the square-cup with a less scattered data than the experimental ones, which can be seen in Figure 11. Superimposing all the experimental and simulation data in a single graph shows the quantitative comparison of the shear angle along a path at the corners. The average shear angle in the experimental results was higher than that in the simulation results and overlapping data between 10 mm and 20 mm were identified at corners 1 and 2 on both sides of the laminate. The gradual increase in the shear angle for the first 20 mm path of the simulation in Figure 11 described the experimental behaviors well. Differences between the forming simulation and experimental results might be caused by a relatively large mesh sizes used in material properties for PAM-FORM simulation to decrease the computation time. The errors were also occurred during the averaging scheme of shear angle used to determine the value for irregular deformation patterns.

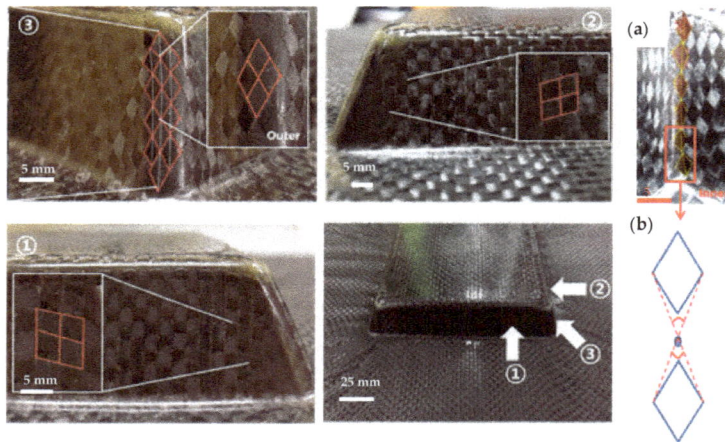

Figure 9. Shear-dominated regions of the PW carbon prepreg at ①, ② and ③, (**a**) Shear angle measurement at the inner corner of ③, (**b**) Averaging the shear angle by connecting imaginary lines between two rhombic shapes.

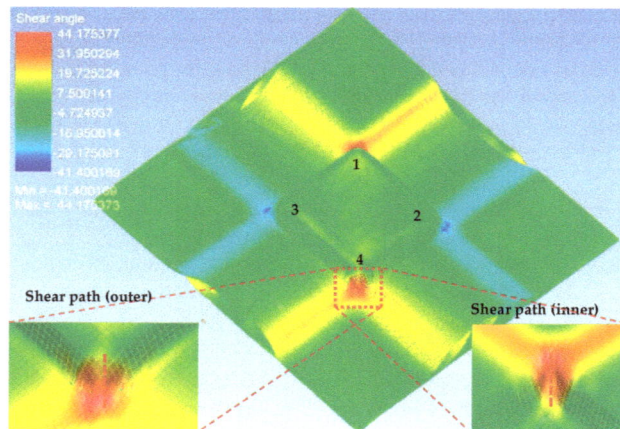

Figure 10. Predicted shear angles in the Type 1 mold design with the pattern 1 prepreg laminate at corner 1, corner 2, corner 3, corner 4 using the PAM-FORM simulation (From top view).

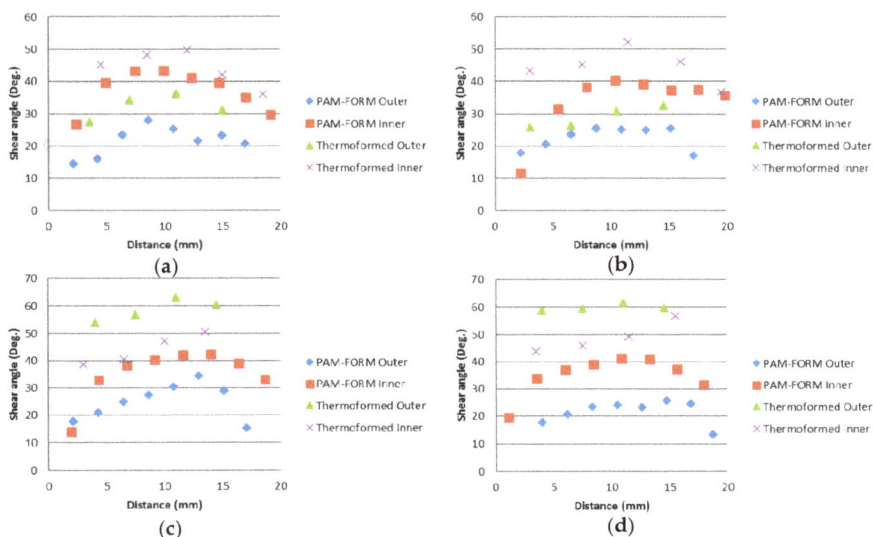

Figure 11. Comparison between the experimental and simulation results for the shear angle at (a) corner 1; (b) corner 2; (c) corner 3; (d) corner 4 as illustrated in Figure 10.

4. Conclusions

The effects of the structural parameters on the manufacturability of fast-cured epoxy resin/carbon prepregs were investigated to optimize the thermoforming process for a mass production line in the automotive industry. The thermoforming process conditions were determined by analyzing the properties of the epoxy resin using differential scanning calorimetry (DSC) and a rheometer. The lowest epoxy viscosity was maintained at 100 °C for 20 min. Therefore, we decided that the best temperature condition at the thermoforming stage was 100 °C, which was then applied to high-temperature mechanical tests. The development of test methods to measure the non-linear mechanical behaviors, such as tensile, in-plane shear (intraply) and frictional force, within the prepregs (interply) for UD carbon, PW carbon and PW glass were introduced. It was expected that a more realistic and predictive virtual forming simulation would be obtained using the input data from non-linear mechanical tests. Two different square-cup shaped molds, type 1 with a tilting angle of 110° and depth of 20 mm and type 2 with a tilting angle of 90° and depth of 40 mm, were prepared to reproduce the shape of a vehicle's roof panel on a small scale. The final product of the thermoformed square-cup proved that the formability and demoldability of the prepreg are affected by the geometry of the mold. Two different stacking sequences of the prepreg laminates were used to assess the effect of formability on the extreme curvature parts, which are the most vulnerable areas to form. Experimental and PAM-FORM simulation data for the shear deformations at each corner were compared to confirm the similarity of the overall tendency of the shear angle in the first 20 mm of the path at each corner.

Author Contributions: Wonoh Lee and Dong Gi Seong designed the experiments; Daeryeong Bae and Shino Kim performed the experiments; Moon Kwang Um and Jin Woo Yi analyzed the data; Daeryeong Bae and Dong Gi Seong wrote the manuscript; All authors discussed and commented on the experiments and manuscript.

Funding: This research was funded by the National Research Foundation (NRF) of Korea, a grant funded by the Ministry of Science and ICT (2018R1A2A2A15020973) and Pusan National University Research Grant, 2017.

Conflicts of Interest: The authors declare no conflict of interest.

References

1. International Council on Clean Transportation. Global Passenger Vehicle Standards. 2014. Available online: http://theicct.org/info-tools/global-passenger-vehicle-standards (accessed on 15 December 2016).
2. Sherwood, J.A.; Fetfatsidis, K.A.; Gorczyca, J.L.; Berger, L. Fabric thermostamping in polymer matrix composite. In *Manufacturing Techniques for Polymer Matrix Composites (PMCs)*; Advani, S.G., Hsiao, K.-T., Eds.; Elsevier: New York, NY, USA, 2012; pp. 139–179.
3. Knibbs, R.H.; Morris, J.B. The effects of fibre orientation on the physical properties of composites. *Composites* **1974**, *5*, 209–218. [CrossRef]
4. Fuller, J.D.; Wisnom, M.R. Exploration of the potential for pseudo-ductility in thin ply CFRP angle-ply laminates via an analytical method. *Compos. Sci. Technol.* **2015**, *112*, 8–15. [CrossRef]
5. Wang, P.; Hamila, N.; Boisse, P. Thermoforming simulation of multilayer composites with continuous fibres and thermoplastic matrix. *Compos. Part B* **2013**, *52*, 127–136. [CrossRef]
6. Sorrentino, L.; Bellini, C. Potentiality of Hot Drape Forming to produce complex shape parts in composite material. *Int. J. Adv. Manuf. Technol.* **2016**, *85*, 945–954. [CrossRef]
7. Wang, P.; Hamila, N.; Pineau, P.; Boisse, P. Thermomechanical analysis of thermoplastic composite prepregs using bias-extension test. *J. Thermoplast. Compos. Mater.* **2014**, *27*, 679–698. [CrossRef]
8. Lee, W.; Padvoiskis, J.; Cao, J.; de Luycker, E.; Boisse, P.; Morestin, F.; Chen, J.; Sherwood, J. Bias-extension of woven composite fabrics. *Int J Mater Form.* **2008**, *1*, 895–898. [CrossRef]
9. Guzman, E.; Hamila, N.; Boisse, P. Thermomechanical analysis, modelling and simulation of the forming of pre-impregnated thermoplastics composites. *Compos. Part A* **2014**, *78*, 211–222. [CrossRef]
10. Gorczyca, J. A Study of the Frictional Behavior of a Plain-Weave Fabric during the Thermostamping Process. Ph.D. Thesis, Department of Mechanical Engineering, University of Massachusetts Lowell, Lowell, MA, USA, January 2004.
11. Ersoy, N.; Potter, K.; Wisnom, M.R.; Clegg, M.J. An experimental method to study the frictional processes during composites manufacturing. *Compos. Part A* **2005**, *36*, 1536–1544. [CrossRef]
12. Thije, R.H.W.; Akkerman, R.; Ubbink, M.; van der Meer, L. A lubrication approach to friction in thermoplastic composites forming processes. *Compos. Part A* **2011**, *42*, 950–960. [CrossRef]
13. Zhang, W.; Ren, H.; Lu, J.; Zhang, Z.; Su, L.; Wang, J.; Zeng, D.; Su, X.; Cao, J. Experimental Methods to Characterize the Woven Composite Prepreg Behavior during the Preforming Process. In Proceedings of the Thirty-First Technical Conference of the American Society for Composites 2016, Williamsburg, VA, USA, 19–21 September 2016.
14. Cao, J.; Akkerman, R.; Boisse, P.; Chen, J.; Cheng, H.S.; de Graaf, E.F.; Gorczyca, J.L.; Harrison, P.; Hivet, G.; Launay, J.; et al. Characterization of mechanical behavior of woven fabrics: Experimental methods and benchmark results. *Compos. Part A* **2008**, *39*, 1037–1053. [CrossRef]
15. Groh, F.; Kappel, F.; Hühne, C.; Brymerski, W. Experimental Investigation of Process Induced Deformations of Automotive Composites with Focus on Fast Curing Epoxy Resins. In Proceedings of the 20th International Conference on Comosite Materials (ICCM-20), Copenhagen, Denmark, 19–24 July 2015.
16. Geissberger, R.; Maldonado, J.; Bahamonde, N.; Keller, A.; Dransfeld, C.; Masania, K. Rheological modelling of thermoset composite processing. *Compos. Part B* **2017**, *124*, 182–189. [CrossRef]
17. Seong, D.G.; Kim, S.; Um, M.K.; Song, S.S. Flow-induced deformation of unidirectional carbon fiber preform during the mold filling stage in liquid composite molding process. *J. Compos. Mater.* **2017**, *52*, 1265–1277. [CrossRef]
18. Boisse, P. *Simulations of Composite Reinforcement Forming, Woven Fabric Engineering*; Dubrovski, P.D., Ed.; InTech: Rijeka, Croatia, 2010; ISBN 978-9-53-307194-7. Available online: https://hal.archives-ouvertes.fr/hal-01635297/document (accessed on 26 October 2016).
19. Xue, P.; Peng, X.; Cao, J. A non-orthogonal constitutive model for characterizing woven composites. *Compos. Part A* **2003**, *34*, 183–193. [CrossRef]
20. Liang, B.; Hamila, N.; Peillon, M.; Boisse, P. Analysis of thermoplastic prepreg bending stiffness during manufacturing and of its influence on wrinkling simulations. *Compos. Part A* **2014**, *67*, 111–122. [CrossRef]
21. Alshahrani, H.; Hojjati, M. A new test method for the characterization of the bending behavior of textile prepregs. *Compos. Part A* **2017**, *97*, 128–140. [CrossRef]

22. Creech, G. Mesoscopic Finite Element Modelling of Non-Crimp Fabrics for Drape and Failure Analyses. Ph.D. Thesis, Cranfield University, Cranfield, UK, 2006.
23. Lee, W.; Um, M.K.; Byun, J.H. Numerical study on thermo-stamping of woven fabric composites based on double-dome stretch forming. *Int. J. Mater. Form.* **2010**, *3*, 1217–1227. [CrossRef]
24. Lee, W.; Cao, J. Numerical simulations on double-dome forming of woven composites using the coupled non-orthogonal constitutive model. *Int. J. Mater. Form.* **2009**, *2*, 145–148.
25. Cartwright, B.K.; de Luca, P.; Wang, J.; Stellbrink, K.; Paton, R. Some proposed experimental tests for use in finite element simulation of composite forming. In Proceedings of the 12th International Conference on Composite Materials (ICCM-12), Paris, France, 5–9 July 1999.
26. Dusi, M.R.; Lee, W.I.; Ciriscioli, P.R.; Springer, G.S. Cure Kinetics and Viscosity of Fiberite 976 Resin. *J. Compos. Mater.* **1987**, *21*, 243–261. [CrossRef]
27. Ersoy, N.; Garstka, T.; Potter, K.; Wisnom, M.; Porter, D.; Stringer, G. Modelling of the spring-in phenomenon in curved parts made of a thermosetting composite. *Compos. Part A* **2010**, *41*, 410–418. [CrossRef]
28. Lightfoot, J.; Wisnom, M.; Potter, K. Defects in woven preforms: Formation mechanisms and the effects of laminate design and layup protocol. *Compos. Part A* **2013**, *51*, 99–107. [CrossRef]
29. Launay, J.; Hivet, G.; Duong, A.; Boisse, P. Experimental analysis of the influence of tensions on in plane shear behaviour of woven composite reinforcements. *Compos. Sci. Technol.* **2008**, *68*, 506–515. [CrossRef]
30. Prodromou, A.; Chen, J. On the relationship between shear angle and wrinkling of textile composite preforms. *Compos. Part A* **1997**, *28*, 491–503. [CrossRef]
31. Wang, P.; Legrand, X.; Boisse, P.; Hamila, N.; Soulat, D. Experimental and numerical analyses of manufacturing process of a composite square box part: Comparison between textile reinforcement forming and surface 3D weaving. *Compos. Part B* **2015**, *78*, 26–34. [CrossRef]

materials

Article

A Comparison of Ethylene-Tar-Derived Isotropic Pitches Prepared by Air Blowing and Nitrogen Distillation Methods and Their Carbon Fibers

Kui Shi [1,2], Jianxiao Yang [1,2,*], Chong Ye [1,2], Hongbo Liu [1,2] and Xuanke Li [1,2,*]

1   College of Materials Science and Engineering, Hunan University, Changsha 410082, Hunan, China; skhnu123@163.com (K.S.); 15874950624@163.com (C.Y.); hndxlhb@163.com (H.L.)
2   Hunan Province Key Laboratory for Advanced Carbon Materials and Applied Technology, Hunan University, Changsha 410082, China
*   Correspondence: yangjianxiao@hnu.edu.cn (J.Y.); xuankeli@hnu.edu.cn (X.L.); Tel.: +86-1521-100-5929 (J.Y.); +86-1599-427-9703 (X.L.)

Received: 7 December 2018; Accepted: 15 January 2019; Published: 18 January 2019

Abstract: Two isotropic pitches were prepared by air blowing and nitrogen distillation methods using ethylene tar (ET) as a raw material. The corresponding carbon fibers were obtained through conventional melt spinning, stabilization, and carbonization. The structures and properties of the resultant pitches and fibers were characterized, and their differences were examined. The results showed that the introduction of oxygen by the air blowing method could quickly increase the yield and the softening point of the pitch. Moreover, the air-blown pitch (ABP) was composed of aromatic molecules with linear methylene chains, while the nitrogen-distilled pitch (NDP) mainly contained polycondensed aromatic rings. This is because the oxygen-containing functional groups in the ABP could impede ordered stack of pitch molecules and led to a methylene bridge structure instead of an aromatic condensed structure as in the NDP. Meanwhile, the spinnability of the ABP did not decrease even though it contained 2.31 wt % oxygen. In contrast, the ABP had narrower molecular weight distribution, which contributed to better stabilization properties and higher tensile strength of the carbon fiber. The tensile strength of carbon fibers from the ABP reached 860 MPa with fiber diameter of about 10 μm, which was higher than the tensile strength of 640 MPa for the NDP-derived carbon fibers.

Keywords: carbon fiber; ethylene tar; isotropic pitch; air blowing

1. Introduction

Carbon fibers (CFs) are widely used in the military, various industries, and sports because of their high mechanical properties, low density, and good conductive properties [1]. The raw materials of pitch-based CFs are usually coal or petroleum-derived by-products, which are abundant, low cost, and have high carbonization yield [2,3]. Therefore, more and more researchers are pursuing low-cost, pitch-based CFs because of their considerable advantage in price and extensive application prospects in the fields of automobiles, sporting goods, building materials, C/C composites, activated carbon fibers, thermal field, etc. [4–9]. Based on this, a lot of raw materials and preparation methods of pitch precursor have been tried, including bromination and subsequent dehydrobromination of naphtha-cracked oil [3], tailored suitable molecular weight portion from Hyper-coal by methylnaphthalene [10], heat treatment of pyrolyzed fuel oil [11], and so on. Ethylene tar (ET) is the by-product of ethylene production, which is rich in resources, low cost, and has low ash content compared to other precursors, especially coal tar pitch, which usually contains primary quinoline insoluble that needs to be removed first. Accordingly, ET is expected to be an ideal raw material for preparing a spinnable pitch. In addition, atmospheric distillation and air blowing are the most common methods to prepare a spinnable pitch for CF production [12–16]. Compared to atmospheric distillation, which mainly involves the removal of light components and condensation polymerization of heavy components during the reaction, air blowing is recognized as an effective method to increase the softening point (SP) and the yield of the pitch. This is because oxidation can link molecules by methylene formed through oxidative dehydrogenation of aliphatic side chains or by oxygen-containing functional groups, such as C–O–C and C=O bridge, formed through oxidation of aliphatic side chains [12,15,17]. Consequently, air blowing can also suppress the formation of mesophase to obtain a homogenous pitch [18]. There has been some research on air blowing of pitches, particularly focusing on the influence of different raw materials, such as coal tar pitch, petroleum pitch, and anthracene oil, or different conditions of air blowing on the final properties of air-blown pitches (ABP) [19–21]. However, these works only investigated the pitches and rarely referred to the preparation of CFs. As a matter of fact, when preparing CFs, oxygen is introduced not only during the air blowing process but also during the stabilization process in order to make the pitch fibers infusible and maintain the fiber shape during the carbonization process. The oxygen introduced in these two stages may have different behaviors during subsequent carbonization processes. Indeed, it is important to consider the oxygen that exists in the pitch precursor as it may lead to a different molecular structure, and the corresponding molecular structure of the pitch would then impact the spinnability of pitches as well as the stabilization and carbonization processes of pitch fibers. In this work, we propose that the oxygen introduced in the pitch precursor can improve the mechanical properties of the resultant CFs prepared at low carbonization temperature below 1200 °C because the introduced oxygen in the pitch precursor is more stable than the introduced oxygen from the stabilization process.

Two kinds of ET-derived isotropic pitches were prepared by nitrogen distillation and air blowing methods in order to compare the differences in their molecular structures and clarify the influence of oxygen introduced during the preparation of the pitch precursor or during the stabilization process of pitch fibers on the properties of the corresponding CFs.

2. Materials and Methods

2.1. Materials

ET was supplied from Wuhan Luhua Yueda Chemical Co. Ltd (Wuhan, China). The ET was used as a raw material to prepare the spinnable pitches; ET is completely soluble in toluene.

2.2. Preparation of Spinnable Pitches

The air-blown pitch was prepared by the air blowing method as follows: (1) The ET was distilled at 250 °C in a 2 L stainless steel reactor to remove the light components to obtain a basic pitch. (2) The basic pitch was air blown at 280 °C for 3 h in 3 L/min air atmosphere to attain oxidized pitch. (3) In order to get the spinnable pitch with high softening point (SP), the oxidized pitch was further heat-treated at 350 °C for 4 h in 3 L/min nitrogen atmosphere. For comparison, the nitrogen-distilled pitch (NDP) with almost the same SP was prepared by heating the ET at 380 °C for 5h in 3L/min nitrogen atmosphere.

2.3. Preparation of Carbon Fibers

The prepared NDP and ABP were spun into pitch fibers (PFs) at a temperature equal to their SP +80 °C using a melt-spinning method with a single-hole spinneret (diameter = 0.2 mm, length/diameter = 3). For this procedure, a laboratory spinning apparatus (Huizhong Dingcheng, Chengdu, China) was used with a nitrogen pressure of 0.4 MPa and winding speeds of 300–500 rpm (100 rpm = 60 m/min) to get PFs with different diameters. The spun PFs were stabilized by heating from room temperature to 280 °C at a rate of 0.5 °C/min and then holding at this temperature for 1 h with an air flow of 500 mL/min. Then, the stabilized fibers (SFs) were successively carbonized at

1200 °C for 30 min at a heating rate of 5 °C/min with a nitrogen flow rate of 100 mL/min. The resultant PFs, SFs, and CFs from the NDP and the ABP were labeled as NDP-PF, ABP-PF, NDP-SF, ABP-SF, NDP-CF, and ABP-CF, respectively.

2.4. Characterization of Pitches and Carbon Fibers

The SP of the pitch was determined by a CFT-100EX capillary rheometer (Shimadzu, Kyoto, Japan). The solubility of the pitch in *n*-hexane, toluene, and quinoline was determined using the Soxhlet extraction method (GB/T 26930.5-2011) to obtain *n*-hexane soluble (HS), *n*-hexane insoluble and toluene soluble (HI-TS), toluene insoluble and quinoline soluble (TI-QS), and quinoline insoluble (QI) fractions. Carbon, hydrogen, sulfur, and nitrogen contents were determined according to SN/T 4764-2017 with a Elementar Vario EL III elemental analyzer (Elementar, Langenselbold, Germany). The oxygen content was obtained by the subtraction method (O=100−C−H−N−S). Fourier transform infrared (FT-IR) spectra were obtained using the KBr disc technique (sample/KBr = 1/100) in a Nicolet iS10 FT-IR spectrometer (Thermo Fisher Scientific, Waltham, MA, USA). Each spectrum was an average of 32 scans with a resolution of 4 cm^{-1}. The solution-state ^{13}C nuclear magnetic resonance (^{13}C-NMR) spectra were obtained using a Bruker 600 MHz Advance NMR spectrometer (Bruker, Karlsruhe, Germany). The quantitative ^{13}C-NMR spectra were recorded by dissolving samples in *d*-chloroform (CDCl$_3$) solvent (sample/CDCl$_3$ = 100 mg/1 mL) with tetramethylsilane used as the chemical shift reference. The ^{13}C-NMR spectra were quantitatively analyzed by the ratio of the peak integral area. X-ray diffraction (XRD) analyses were performed by a D8 Advance diffractometer (Bruker, Karlsruhe, Germany) with Cu Kα radiation (λ = 0.15406 nm) generated at 32 kV and 50 mA with a scan speed of 1 °/min for 2 theta values between 5° and 80°. The thermogravimetric (TG) properties of samples were measured using a STA 449 F5 thermal analyzer (Netzsch, Selb, Germany). The thermal stability and coking value (CV) of the obtained pitches were analyzed in 40 mL/min nitrogen atmosphere with a heating rate of 5 °C/min to 900 °C. Meanwhile, to evaluate the stabilization properties of spun PFs, the obtained PFs were also analyzed by TG with different heating rates (0.5 °C/min, 1 °C/min, 2 °C/min, 4 °C/min) to 600 °C in 40 mL/min air atmosphere to find their maximum weight gain (W$_{max}$) and the corresponding maximum temperature (T$_{max}$) as well as to calculate their reaction activation energy (E$_a$). The gas released during the carbonization of PFs and SFs were measured by Hiden Analytical HAS-301-1474 mass spectrometer (Hiden Analytical, Warrington, UK) coupled with TG, which was heated from 20 to 1200 °C at a rate of 10 °C /min and an argon flow of 20 mL/min. The MS was performed at RGA mode with a secondary electron multiplier, and the quartz capillary connected to the thermal analyzer was heated to 160 °C. The morphologies and the diameter of the CFs were observed by JSM-6700F field emission scanning electron microscope (SEM, JEOL, Tokyo, Japan) with 5 kV. The tensile strength and Young's modulus of CFs were measured at room temperature using monofilaments with a gauge length of 20 mm according to the standard (ASTM D4018-2011). The diameter of CFs was observed by SEM after the tensile experiment. The tensile strength and Young's modulus were evaluated from the mean value of 30 tests, with the values distributing within 10%.

3. Results and Discussion

3.1. Characterization of NDP and ABP

The general characteristics of pitches are summarized in Table 1. The yield of the ABP was higher than that of the NDP, and their SP was almost the same. This result is in line with results from previous researches because the air blowing method can increase SP more quickly [17]. It was apparent that air blowing brought large amounts of oxygen into the pitch as the oxygen content of the ABP was up to 2.31%. This was higher than the oxygen content of the NDP, which was 0.78%. Compared to the ET, the solubility of the NDP and the ABP in *n*-hexane and toluene decreased; HI-TS increased from 48.4% for the ET to 59.8% and 83.2% for the NDP and the ABP, respectively, as shown in Figure 1. It should be noted that TI appeared in both the NDP and the ABP, but the TI of the NDP was larger than that of the ABP. In addition, QI appeared in the NDP, which demonstrated that the nitrogen distillation method was more effective in accelerating the polymerization reaction and forming larger molecules than the air blowing method [22]. This could also be confirmed by the larger C/H atom rate of the NDP compared to the ABP, i.e., 1.40 and 1.26, respectively. Although the NDP had larger molecules, it is believed that the molecular weight distribution of the ABP was narrower and more uniform than the NDP due to its quite high HI-TS content. The SP of ABP was equal to that of the NDP, even at lower treatment temperature, because larger molecules were produced through oxidative cross-linking of small molecules when air blowing. Meanwhile, oxidative cross-linking suppressed molecule aggregation and uneven polymerization that might happen in the nitrogen distillation process. This more homogeneous composition of ABP would be beneficial for its spinning performance.

Table 1. Softening point (SP), pitch yield, and elemental analysis results of the air-blown pitch (ABP) and the nitrogen-distilled pitch (NDP).

Sample	SP (°C)	Yield (%)	Elemental Analysis (%)					
			C	H	N	S	O	C/H
NDP	253	22	93.53	5.56	0.02	0.04	0.78	1.40
ABP	252	28	91.55	6.06	0.05	0.10	2.31	1.26

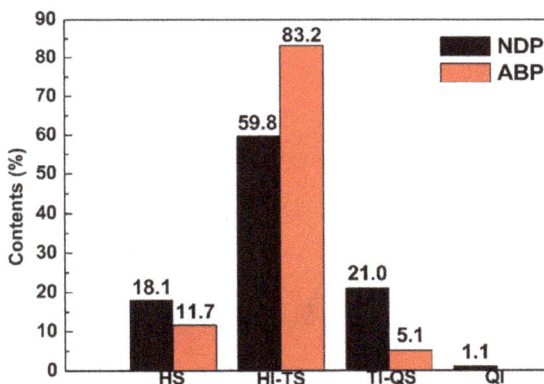

Figure 1. Solubility parameter (*n*-hexane soluble (HS), *n*-hexane insoluble and toluene soluble (HI-TS), toluene insoluble and quinoline soluble (TI-QS), and quinoline insoluble (QI)) of the NDP and the ABP.

FT-IR analyses, shown in Figure 2, were carried out to verify the functional groups in the NDP and the ABP. The absorption peaks at 3050 cm^{-1} and 1600 cm^{-1} were assigned to the presence of aromatic C–H and aromatic C–C stretching vibrations, respectively. The stronger absorbance at 2920 cm^{-1} and 2850 cm^{-1} in the ABP corresponded to methylene hydrogen asymmetric and symmetric stretching vibrations, respectively [23]. The peak at 1450 cm^{-1} corresponding to methylene hydrogen bending vibration was clearly strong in intensity. The broad peak at 3300–3600 cm^{-1} might be ascribed to hydroxyl of H_2O in the pitches [24]. It must also be mentioned that a new peak appeared at 1700 cm^{-1} for the ABP compared to the NDP, which was assigned to the C=O stretching vibration. This could be easily explained by the higher oxygen content of the ABP, as shown in Table 1. Another weak bond at 1260 cm^{-1} belonging to the C–O stretching vibration could also be observed. The emergence of oxygen-containing functional groups indicated that oxygen might connect pitch molecules as an oxygen bridge by air blowing.

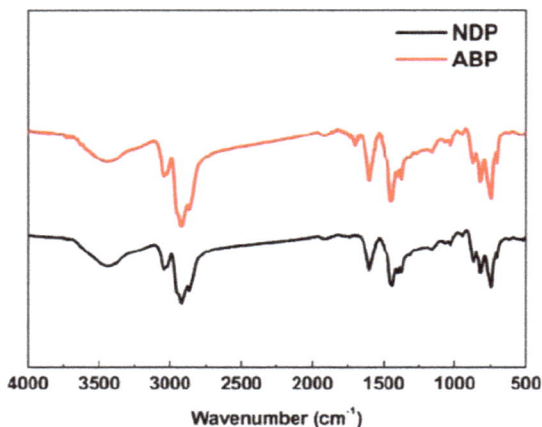

Figure 2. FT-IR spectra of the NDP and the ABP.

In order to further study the molecular structure of the NDP and the ABP, ^{13}C-NMR was performed. The ^{13}C-NMR spectra of the NDP and the ABP are plotted in Figure 3. The normalized integration data of ^{13}C-NMR spectra is presented in Table 2. The larger aromaticity of the NDP could be demonstrated by larger C_{ar}/C_{al} compared to the ABP, i.e., 5.29 and 2.85, respectively. Most of the aromatic carbon in NDP was $C_{ar1,3}$, suggesting abundant pericondensed structure in the NDP [25]. This result indicated more condensed large molecules in the NDP produced by heat treatment, which was reflected by the higher insoluble content, as shown in Figure 1. More aliphatic carbon in the form of CH_2 and $C_{\alpha2}$ in the ABP compared to the NDP indicated more methylene bridge structures in the ABP, formed by oxidative dehydrogenation, which is in accordance with the above analysis.

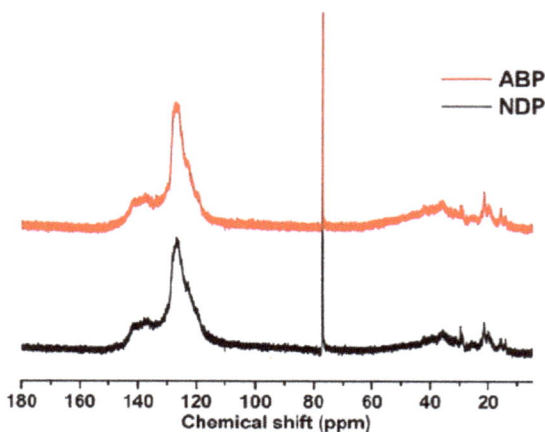

Figure 3. ^{13}C-NMR spectra of the NDP and the ABP.

Table 2. Normalized integration data based on the ^{13}C NMR spectra.

Samples	C_{al}			C_{ar}		C_{al}	C_{ar}	C_{ar}/C_{al}
	CH_3	CH_2	$C_{\alpha2}$	$C_{ar1,3}$	$C_{ar1,2}$			
NDP	0.04	0.08	0.04	0.60	0.24	0.16	0.84	5.29
ABP	0.03	0.14	0.09	0.52	0.22	0.26	0.74	2.85

CH_3, methyl carbon; CH_2, methylene (CH_2) carbon α or further from an aromatic ring in free side chain; $C_{\alpha2}$, CH_2 carbon in bridge/hydroaromatic structures; $C_{ar1,3}$, pericondensed aromatic carbon (C_{ar3}) and protonated aromatic carbon (CH_{ar}); $C_{ar1,2}$, catacondensed aromatic carbon, aromatic carbon with both heteroatomic or aromatic substituents (C_{ar2}), and the region correspondent to aromatic carbon joined to aliphatic chains; C_{al}, total aliphatic carbon; C_{ar}, total aromatic carbon.

Figure 4 shows the XRD graphs of the NDP and the ABP. Both of them showed broad peaks between 10° and 30°. Peaks between 10° and 20° were attributable to asphaltene components, while those between 20° and 30° were attributable to stacked molecular structures [26]. Therefore, the stacked structure was more evident in the NDP than that in the ABP due to larger condensed aromatic molecules in the NDP, as shown in Figure 1. In contrast, the oxygen in the ABP impeded ordered stack of molecules. Nevertheless, more symmetrical peak of ABP may indicate more uniform composition, which is consistent with the extraordinarily uniform components presented in Figure 1.

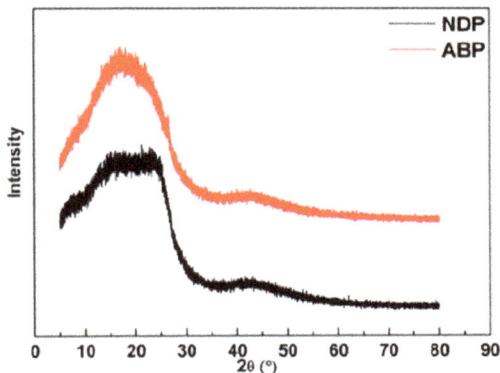

Figure 4. X-ray diffractograms of the NDP and the ABP.

3.2. Spinning Properties of NDP and ABP

The viscosity–temperature curves of NDP and ABP are shown in Figure 5. The viscosity of both the NDP and the ABP decreased sharply with increasing temperature when the temperature was lower than 310 °C, then decreased gradually to about 330 °C, while the viscosity–temperature curves became almost flat above 330 °C because of the temperature sensitive property of the pitch. It can be seen that the viscosity–temperature curve of the ABP showed two jumping steps between 310 °C and 330 °C, which suggested that the ABP had worse spinning performance than the NDP. However, both of them had excellent spinning performance when the NDP and the ABP were spun into PFs by the melt-spinning method at 335 °C. Therefore, the results indicated that the spinnability of the ABP had not deteriorated even though it had more oxygen, which is usually considered as an impurity atom. This may be attributed to the homogenous components of the ABP, as previously mentioned.

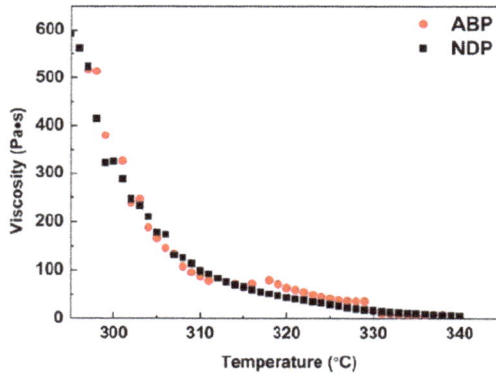

Figure 5. Viscosity–temperature curves of the NDP and the ABP.

3.3. Stabilization and Carbonization of NDP-PF and ABP-PF

Stabilization is a crucial process to determine the properties of CFs due to the formation of intermolecular cross-linking in an oxidizing atmosphere, which can ensure that the shape of fibers do not change in subsequent carbonization processes [27]. Therefore, proper stabilization parameter should be adopted. In order to optimize the stabilization process, TG was used to measure the oxidation reactivity of PFs, and the results are presented in Figure 6. The lower initial temperature of weight gain indicated higher oxidation reactivity of ABP-PF compared to NDP-PF. This was also proven by the lower E_a and T_{max} of ABP-PF, as shown in Table 3. The higher oxidation reactivity of ABP-PF might have been due to the lightest components remaining in the ABP instead of being removed by distillation at high temperature as in the NDP. However, the W_{max} in oxidation of ABP-PF was less than NDP-PF, i.e., 11.6% and 12.8%, respectively, as exhibited in Table 3. This result could be interpreted as showing that the higher oxygen in the ABP could restrain more oxygen from diffusing into the PFs. This viewpoint can be demonstrated by the similar oxygen content of NDP-SF and ABP-SF, as presented in Table 3.

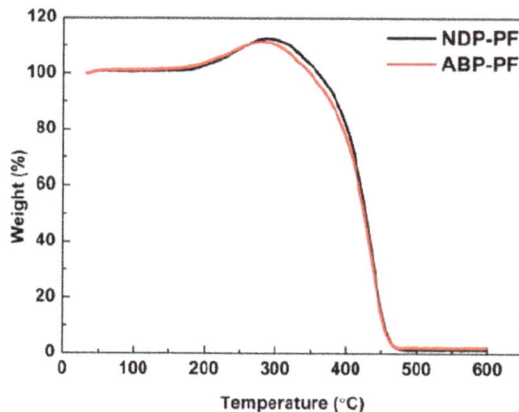

Figure 6. Thermogravimetric (TG) curves of NDP-PF and ABP-PF during stabilization process in air atmosphere.

Table 3. Oxidation and carbonization properties of pitch fibers.

Samples	T_{max} (°C)	W_{max} (%)	E_a (kJ/mol)	Oxygen Content (%)	Yield (%)
NDP-SF	286	12.8	124.5	25.46	114.0
ABP-SF	278	11.6	120.5	25.72	112.5
NDP-CF	-	-	-	8.96	71.3
ABP-CF	-	-	-	10.29	71.0

T_{max}, W_{max}, and E_a were calculated from TG results of NDP-PF and ABP-PF in air atmosphere.

After 1200 °C carbonization, both NDP-CF and ABP-CF had high yield over 70%, which is favorable for producing low-cost, general-purpose CFs. Higher oxygen content of ABP-CF compared to NDP-CF, as shown in Table 3, denoted that less oxygen of ABP-CF was given off during the carbonization process in view of the similar oxygen content of NDP-SF and ABP-SF. Less gas was released in the carbonization process of ABP-SF (35.7%) compared to the NDP-SF (37.7%), which might be beneficial for the mechanical properties of CFs because fewer defects would be generated. This also illustrates that the oxygen existing in the pitch precursor was more stable than that introduced in the stabilization process. This could be verified by the TG–MS results of the CO and CO_2 given off from oxygen-containing groups of fibers during carbonization process for NDP-SF and ABP-PF, as shown in Figure 7a,b. Both CO and CO_2 were released at two stages; the first peak was located at 400–800 °C and 270–850 °C, respectively. Then, the amount of release began to increase rapidly for both CO and CO_2, especially for NDP-SF. It should be noted that the difference in the release amount of CO_2 for NDP-SF and ABP-PF was more distinct. The NDP-SF revealed higher peak intensity and magnitude of CO_2 in both stages, especially the high temperature stage, compared to ABP-PF. This showed that the removal of oxygen was more arduous below 1200 °C when the oxygen was from the pitch precursor introduced by air blowing than the oxygen from the stabilization process. Therefore, moderate amounts of oxygen in the pitch precursor could improve the stabilization and carbonization properties of PFs.

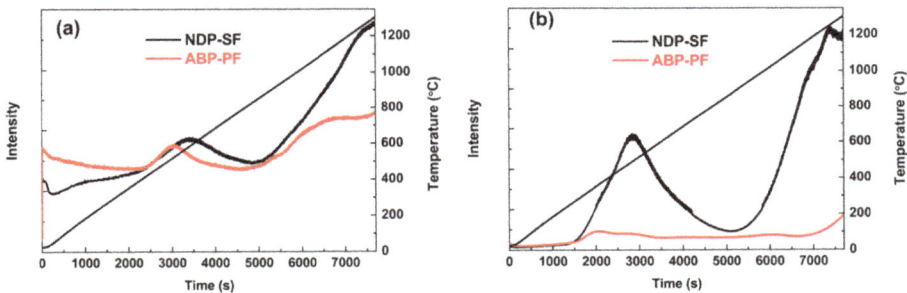

Figure 7. MS curves of NDP-SF and ABP-PF during carbonization process in argon (**a**) CO and (**b**) CO_2.

3.4. Morphology and Mechanical Properties of NDP-CF and ABP-CF

The SEM micrographs of CFs are shown in Figure 8. ABP-CF expressed more homogeneous and smooth surfaces and cross sections compared to NDP-CF, corresponding to more homogeneous precursor pitch. Moreover, the CFs exhibited no fusing, indicating that the stabilization process had been adequate. The cross section of the CFs showed a glass-like fracture surface, indicating that the precursor pitch was isotropic. The tensile strength and Young's modulus of CFs are presented in Figure 9. ABP-CF had a tensile strength of 550–895 MPa and modulus of 31–45 GPa when the diameter decreased from 14.1 to 9.8 µm; these figures were consistently higher than for NDP-CF, which had tensile strength of 448–748 MPa and modulus of 27–35 GPa as the diameter decreased from 13.1 to 8.7 µm. In both cases, the tensile strength decreased with increasing diameter. The mechanical properties reached the level of commercial, general-purpose CFs. Higher tensile strength can be attributed to the uniform composition of the ABP and less gas released in the carbonization process, which can produce defects in CFs due to the existence of oxygen, as previously discussed. Whether the greater amount of oxygen, which would be removed at higher temperature above 1200 °C had an unfavorable effect on the properties of CFs needs further investigation.

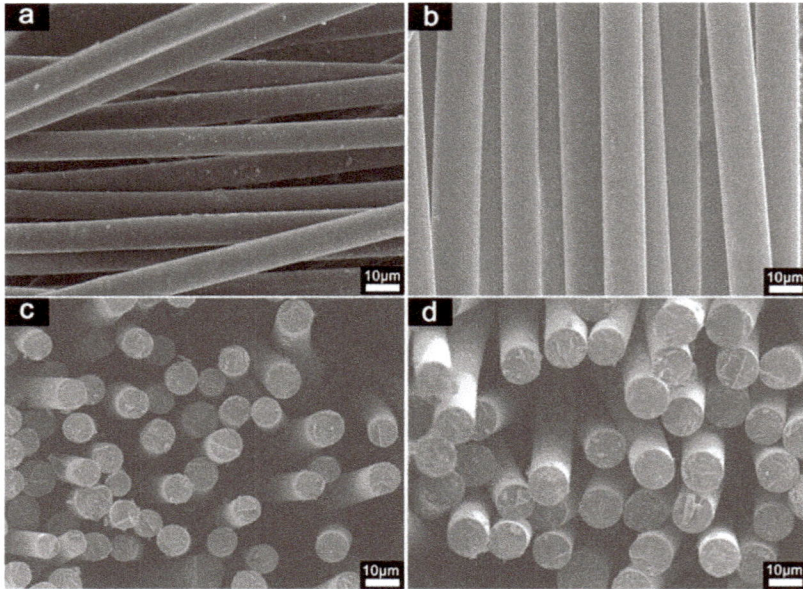

Figure 8. SEM micrographs of (**a,c**) NDP-CF and (**b,d**) ABP-CF.

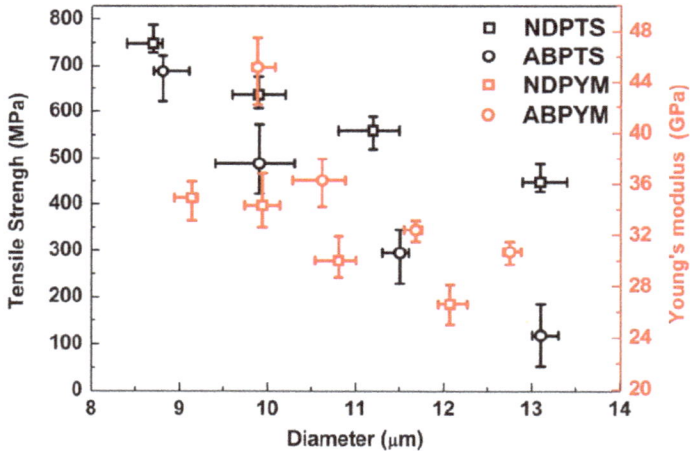

Figure 9. The mechanical properties of NDP-CF and ABP-CF.

4. Conclusions

Two isotropic pitches were prepared through atmospheric distillation and air blowing, respectively. The ABP had a higher yield and more homogeneous composition than the NDP because oxygen could connect smaller molecules and suppress the ordered stack of larger molecules. The oxygen existing in the pitch precursor was harder to remove during carbonization, contributing to higher tensile strength

Materials **2019**, *12*, 305

of ABP-CF compared to NDP-CF. The tensile strengths of ABF-CF and NDP-CF were 860 MPa and 640 MPa, respectively, when the diameter was about 10 μm, and the modulus was 41 GPa and 33 GPa, respectively. The results revealed that the oxygen introduced during the pitch precursor preparation process made a significant difference compared to the PF stabilization process. The evolution of introduced oxygen during the pitch precursor, stabilization, and carbonization processes will be investigated in detail in future work.

Author Contributions: Conceptualization, J.Y. and X.L.; Methodology, K.S.; Validation, K.S., J.Y., C.Y., H.L., and K.S.; Formal analysis, K.S. and J.Y.; Investigation, K.S.; Data curation, K.S. J.Y., and X.L.; Writing—original draft preparation, K.S.; Writing—review and editing, J.Y. and X.L.; Supervision, H.L.; Project administration, X.L.; Funding acquisition, J.Y. and X.L.

Funding: This research was funded by (i) the Key Project of Chinese National Programs (2016YFB0101702), (ii) the National Science Foundation for Young Scientists of China (Grant No. 51702094), and (iii) the Natural Science Foundation for Young Scientists of Hunan Province, China (Grant No. 2017JJ3014).

Acknowledgments: The authors appreciate the support of Wuhan Luhua Yueda Chemical Co. Ltd, China, for providing the ET.

Conflicts of Interest: The authors declare no conflict of interest.

References

1. Frank, E.; Steudle, L.M.; Ingildeev, D. Carbon Fibers: Precursor Systems, Processing, Structure, and Properties. *Angew. Chem. Int. Ed.* **2014**, *53*, 5262–5298. [CrossRef] [PubMed]
2. Mochida, I.; Toshima, H.; Korai, Y. Blending mesophase pitch to improve its properties as a precursor for carbon fibre I. Blending of PVC pitch into coal tar and petroleum-derived mesophase pitches. *J. Mater. Sci.* **1988**, *23*, 670–677. [CrossRef]
3. Kim, B.; Eom, Y.; Kato, O. Preparation of carbon fibers with excellent mechanical properties from isotropic pitches. *Carbon* **2014**, *77*, 747–755. [CrossRef]
4. Mora, E.; Blanco, C.; Prada, V. A study of pitch-based precursors for general purpose carbon fibers. *Carbon* **2002**, *40*, 2719–2725. [CrossRef]
5. Xu, Y.; Chung, D.D.L. Silane-treated carbon fiber for reinforcing cement. *Carbon* **2001**, *39*, 1995–2001. [CrossRef]
6. Fu, X.; Lu, W.; Chung, D.D.L. Ozone treatment of carbon fiber for reinforcing cement. *Carbon* **1998**, *36*, 1337–1345. [CrossRef]
7. Chand, S. Review carbon fibers for composites. *J. Mater. Sci.* **2000**, *35*, 1303–1313. [CrossRef]
8. Shirvanimoghaddam, K.; Hamim, S.U.; Karbalaei, A.M. Carbon fiber reinforced metal matrix composites: Fabrication processes and properties. *Compos. Part A-Appl. Sci.* **2017**, *92*, 70–96. [CrossRef]
9. Park, S.H.; Kim, C.; Choi, Y.O. Preparations of pitch-based CF/ACF webs by electrospinning. *Carbon* **2013**, *41*, 2655–2657. [CrossRef]
10. Yang, J.; Nakabayashi, K.; Miyawaki, J. Preparation of pitch based carbon fibers using Hyper-coal as a raw material. *Carbon* **2016**, *106*, 28–36. [CrossRef]
11. Kim, J.G.; Kim, J.H.; Song, B. Synthesis and its characterization of pitch from pyrolyzed fuel oil (PFO). *J. Ind. Eng. Chem.* **2016**, *36*, 293–297. [CrossRef]
12. Barr, J.B.; Lewis, I.C. Chemical changes during the mild air oxidation of pitch. *Carbon* **1978**, *16*, 439–444. [CrossRef]
13. Mochida, I.; Inaba, T.; Korai, Y. Carbonization properties of carbonaceous substances oxidized by air blowing—I: Carbonization behaviors and chemical structure of residual oils oxidized by air blowing. *Carbon* **1983**, *21*, 543–552. [CrossRef]
14. Mochida, I.; Inaba, T.; Korai, Y. Carbonization properties of carbonaceous substances oxidized by air blowing—II: Acid-catalyzed modification of oxidized residual oil for better anisotropic development. *Carbon* **1983**, *21*, 553–558. [CrossRef]
15. Blanco, C.; Santamar, A.R.; Bermejo, J. A comparative study of air-blown and thermally treated coal-tar pitches. *Carbon* **2000**, *38*, 517–523. [CrossRef]
16. Menendez, R.; Blanco, C.; Santamaria, R. On the chemical composition of thermally treated coal-tar pitches. *Energy Fuels* **2001**, *15*, 214–223. [CrossRef]

17. Prada, V.; Granda, M.; Bermejo, J. Preparation of novel pitches by tar air-blowing. *Carbon* **1999**, *37*, 97–106. [CrossRef]

18. Maeda, T.; Zeng, S.M.; Tokumitsu, K. Preparation of isotropic pitch precursors for general purpose carbon fibers (GPCF) by air blowing—I. Preparation of spinnable isotropic pitch precursor from coal tar by air blowing. *Carbon* **1993**, *31*, 407–412. [CrossRef]

19. Zeng, S.M.; Maeda, T.; Tokumitsu, K. Preparation of isotropic pitch precursors for general purpose carbon fibers (GPCF) by air blowing—II. Air blowing of coal tar, hydrogenated coal tar, and petroleum pitches. *Carbon* **1993**, *31*, 413–419. [CrossRef]

20. Fernández, J.J.; Figueiras, A.; Granda, M. Modification of coal-tar pitch by air-blowing—I. Variation of pitch composition and properties. *Carbon* **1995**, *33*, 295–307. [CrossRef]

21. Fernández, A.L.; Granda, M.; Bermejo, J. Air-blowing of anthracene oil for carbon precursors. *Carbon* **2000**, *38*, 1315–1322. [CrossRef]

22. Mishra, A.; Saha, M.; Bhatia, G. A comparative study on the development of pitch precursor for general-purpose carbon fibres. *J. Mater. Process. Tech.* **2005**, *168*, 316–320. [CrossRef]

23. Wu, B.; Hu, H.; Zhao, Y. XPS analysis and combustibility of residues from two coals extraction with sub- and supercritical water. *J. Fuel. Chem. Technol.* **2009**, *37*, 385–392. [CrossRef]

24. Matsumoto, T.; Mochida, I. A structural study on oxidative stabilization of mesophase pitch fibers derived from coal tar. *Carbon* **1992**, *30*, 1041–1046. [CrossRef]

25. Andersen, S.I.; Jensen, J.O.; Speight, J.G. X-ray Diffraction of subfractions of petroleum asphaltenes. *Energy Fuels* **2005**, *19*, 2371–2377. [CrossRef]

26. Korai, Y.; Mochida, I. Molecular assembly of mesophase and isotropic pitches at their fused states. *Carbon* **1992**, *30*, 1019–1024. [CrossRef]

27. Drbohlav, J.; Stevenson, W.T.K. The oxidative stabilization and carbonization of a synthetic mesophase pitch, part I: The oxidative stabilization process. *Carbon* **1995**, *33*, 693–711. [CrossRef]

materials

MDPI

Article

Evaluation of an FE Model for the Design of a Complex Thin-Wall CFRP Structure for a Scientific Instrument

Enrique Casarejos *, Jose C. Riol †, Jose A. Lopez-Campos, Abraham Segade and Jose A. Vilan ‡

Department Mechanical Engineering, University of Vigo, E-36310 Vigo, Spain; jriol@uvigo.es (J.C.R.); joseangellopezcampos@uvigo.es (J.A.L.-C.); asegade@uvigo.es (A.S.); jvilan@uvigo.es (J.A.V.)
* Correspondence: e.casarejos@uvigo.es; Tel.: +34-986-813-780
† This work is part of his PhD Thesis.
‡ Currently retired.

Received: 11 December 2018; Accepted: 1 February 2019; Published: 5 February 2019

Abstract: In this paper, the reliability of a finite element (FE) model including carbon-fibre reinforced plastics (CFRPs) is evaluated for a case of a complex thin-wall honeycomb structure designed for a scientific instrument, such as a calorimeter. Mechanical calculations were performed using FE models including CFRPs, which required a specific definition to describe the micro-mechanical behaviour of the orthotropic materials coupled to homogeneous ones. There are well-known commercial software packages used as powerful tools for analyzing structures; however, for complex (many-parts) structures, the models become largely time consuming for both definition and calculation, which limits the appropriate feedback for the structure's design. This study introduces a method to reduce a highly nonlinear model, including CFRPs, into a robust, simplified and realistic FE model capable of describing the deformations of the structure with known uncertainties. Therefore, to calculate the deviations of our model, displacement measurements in a reduced mechanical setup were performed, and then a variety of FE models were studied with the objective to find the simplest model with reliable results. The approach developed in this work leads to concluding that the deformations evaluated, including the uncertainties, were below the actual production tolerances, which makes the proposed model a successful tool for the designing process. Ultimately, this study serves as a future reference for complex projects requiring intensive mechanical evaluations for designing decisions.

Keywords: composite; CFRP; thin-wall; finite element model; contact problem

---

1. Introduction

The deployment of reinforced plastics for structural components is a continuously growing option for many industrial fields. Either carbon, glass, or kevlar fibre reinforced plastics provide materials with large strength-to-mass ratios. The success of the application of these materials is linked to the efforts for developing models and calculation tools. The specialized literature is ample, e.g., the recent reference books [1,2], papers dedicated to laminated carbon-fibre reinforced plastics (CFRPs) [3], honeycomb cells [4] and large deformations of layered materials [5].

This research presents a study case of a scientific instrument requiring an important mechanical structure. Many types of instruments typically demand light structures with the requisite of being a robust and safe mechanical frame, and usually also providing an accurate and tight positioning of the critical active elements. An ideal instrument should have a maximal active volume free of structural parts since more structural parts have less sensitivity and more additional side effects. Honeycomb like structures, made of thin-wall parts, can provide excellent solutions. Applying to this structure,

CFRP materials may be a very appropriate choice. Instruments with these selections can be found in new generation large telescopes [6], spatial applications [7,8] and particle physics instruments [9–12].

Scientific instrument design requires the analysis of demands from rather different sources. The physics codes must evaluate the adequacy of sensor volume, shape, position, and segmentation for maximum efficiency for the instrument goals. The mechanical design has to be adapted to the sensor needs, and calculations evaluate the structural requirements. The structure deformations influencing the location and integrity of the sensors require re-designing stiffer parts, and finer adjustments of the mechanical system would provide an optimal sensor orientation.

Despite the integration of CFRP pre- and post-processors in commercial software packages dedicated to finite element (FE) models, the definition quality of the orthotropic materials may be limited due to the implementation of the micro-mechanical models. It is also common that many and diverse CFRP fabrics in the market suffer from the scarce experimental data available to correctly describe the material. If the structure is complex and the number of parts is large, any FE model becomes largely time consuming for both its definition and its calculation. In this paper, an effective and robust FE model was used as a design tool, which has been developed and described along the project using a methodology that can be deployed for similar complex structural cases.

With the objective to show the calculation of a complete instrument structure to better understand deformations, and be able to correctly adjust the orientation of the sensors, this paper is organised as follows: in Section 2, the instrument is described. In Section 3, the mechanical test bench is presented to obtain the reference data. In Section 4, the settings used in the FE models implemented for the study are discussed. In Section 5, the results related to the simulation models and experimental setups are compared and the best FE model selected. In Section 6, the best model is applied to the complete structure of the instrument and the results presented, and, finally, in Section 7, the conclusions of this work are stated.

2. The CALIFA-FAIR Calorimeter: A Case Reference

The CALorimeter for the In Flight detection of gamma-rays and light charged pArticles (CALIFA) is an instrument developed for experiments in the international research facility FAIR (Germany), and under R&D [13,14] involving 12 different European institutes.

CALIFA is a detector for nucleus-to-nucleus collisions dedicated to particle physics studies. Surrounding the collision point, it is necessary an adequate segmentation into single sensors to fulfill the required detector sensitivity. Sensors are prismatic bodies made of radiation-sensitive scintillating crystals with lengths (between 170 and 220 mm) and shapes depending on their location, with an average volume of 147 cm^3, and with glued opto-electronic circuit boards at one end. For the correct working of the instrument, it is necessary to define the positions of 1952 individual sensors with a total weight of about 1300 kg, and also to provide the right shape for the instrument.

As mentioned in the Introduction, an adequate honeycomb structure was used as mechanical support [15] to hold the sensors (crystals), to conform to the shape around the centre, and to position the crystals with enough stiffness to guarantee the orientation of all the individual sensors. The honeycomb walls between sensors have to be as thin as possible to reduce to a minimum the interactions of the particles and radiation through the matter. This effect is ruled by the atomic number of the material; therefore, carbon-compounds are always preferred to typical light metals.

Construction of the CALIFA Calorimeter

In this study, a CFRP fabric was used for the construction of the honeycomb structure (*CF-structure*), which was built with 512 CFRP thin-wall parts of 0.3 mm each. Each CF-part was a hollow prismatic box with lengths that range between 210 and 270 mm (Figure 1a,b), and were glued together with the neighbouring parts [15]. In addition, these CF-parts with adapted shapes were distributed in sixteen rings around an opening of 600 mm diameter, each ring built with 32 CF-parts (see Figure 1c). The CF-structure had an envelope of diameter 1060 mm and a length of 990 mm. The weight ratio of the structure with respect to the total mass resulted in a value as low as 0.7%.

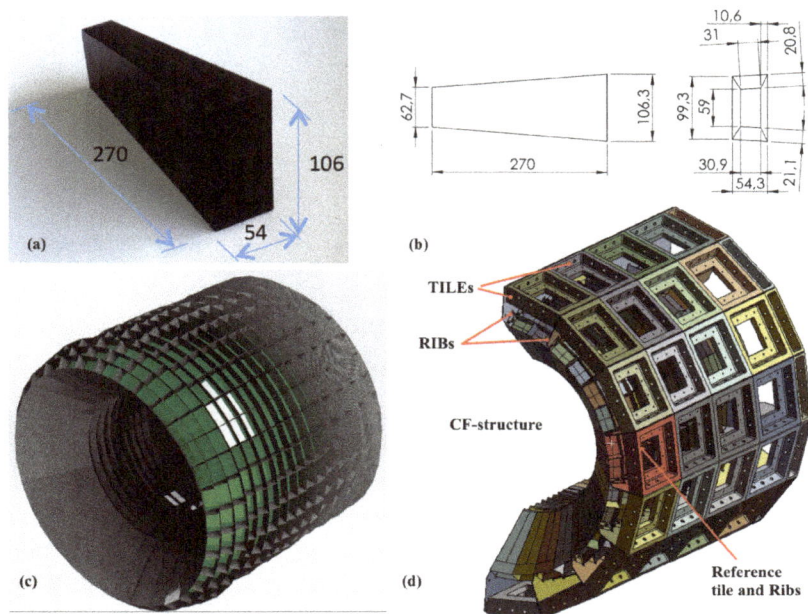

Figure 1. (**a**,**b**) picture and drawing of one CF-part (in mm); (**c**) drawing of the honeycomb CF-structure. The sixteen rings around the opening are built with 32 prismatic parts each; (**d**) cover structure built with 64 tiles for supporting the CFRP structure. Only one half is shown.

Since sensors require a light- and gas-tight enclosure, a cylindrical cover around the CF-structure was the natural shape solution. This cover was designed with 64 similar parts called *tiles*; they were assembled in a regular pattern with an outer diameter of 1200 mm (Figure 1c). There were parts called *ribs* to hold the inner CF-structure as well as to connect with the cover. The ribs were 96 comb-like shaped plates with flaps. The reference tile and ribs for further calculations are marked in red in Figure 1d, and correspond to those in Figure 2. Each pair of ribs grabbed between their outer and inner flaps the walls of the CF-structure (Figure 2a and for more detail Figure 3b). The ribs were fastened in between the tiles of the cover providing the support of the inner CF-structure (Figure 1c). The cover is a very stiff and robust fastened assembly, which can be mounted on a gantry or equivalent external frame structure for the ultimate placement of the instrument [15].

The assembly of the CF-structure contains 512 CF-parts, which require being defined wall by wall in any FE model to specify the orientation of the fibre layer-by-layer of the CFRP because of the orthotropic nature of the CFRP material. Additionally, there are hundreds of metallic parts, and therefore thousands of contacts between the CF-parts, ribs, tiles, fasteners, and sensors inside the structure. This calorimeter was designed to be split in two equal halves by the vertical mid-plane (Figure 1d). This symmetry allows the calculation of only one-half of the structure but does not change the scale of the problem. Since the overall design required the evaluation of many different aspects in a feedback process with Physics simulations, there was a need to develop an effective and reliable FE model for the structure although reducing as much as possible its complexity for both definition and calculation.

A FE model was built as realistic as possible while preserving an effective definition for design. This study was based on selected definitions according to the results of benchmarking tests previously done on a singular part of the structure, and then an evaluation was conducted from a collection of FE models with different material and contact definitions. All the results were crosschecked with the reference data to get the deviation values and to allow the evaluation and analysis of the model performance and reliability.

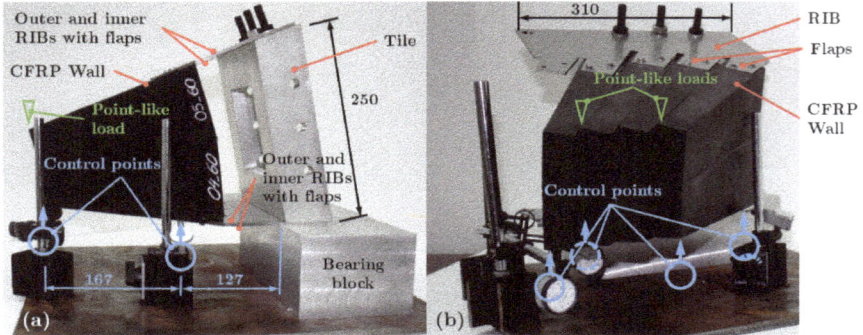

Figure 2. Pictures of the mechanical test bench. (**a**) setup with two CF-parts and two control points; (**b**) setup with eight CF-parts and three control points (one control point on the front was removed for this picture).

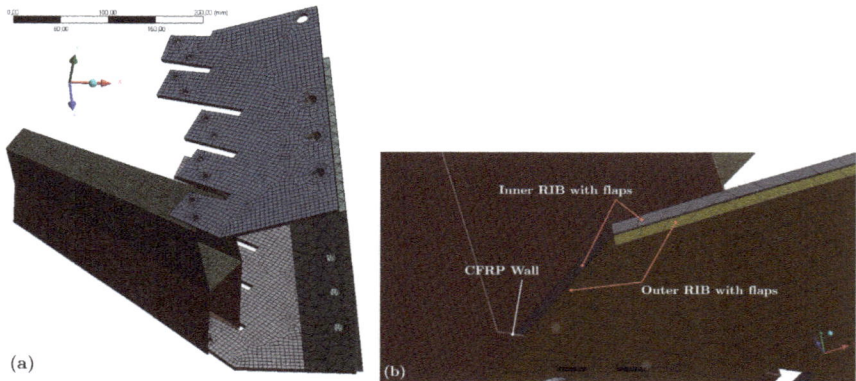

Figure 3. Drawings showing the mesh model used for the FE models. (**a**) setup with two CF-parts; (**b**) a detailed view of the rib-flap region where the CF-part is fastened.

3. Mechanical Tests Setup and Measurements to Obtain the Reference Data

Mechanical tests were performed in a robust setup, providing simple and easy to interpret data to compare the results to the FE models. The setups were mounted with different CF-parts to ensure that the use of specific parts made no difference to the results.

Two different setups were mounted with two and eight CF-parts, respectively, with dimensions as shown in Figure 2. The two configurations provided results with an important scale factor (400%) in order to study the whole structure from these reduced sets. The CF-parts were clamped firmly in between the flaps of the ribs at both sides, and fastened by bolts. The ribs, 2 mm thickness stainless steel, were fastened to a tile and a bearing block with bolts, being the same mounting system as for the whole structure.

Measuring deflection in the most unfavourable direction, where gravity acts on the most perpendicular rib of a tile corresponding to the red color tile in Figure 1d, ensures the appropriate performance of the structure. The lower wall of the CF-parts was set in the horizontal plane.

Dial gauges (*Mitutoyo Corporation*, Kawasaki, Japan, 0.005 mm resolution) were firmly placed at the base plate to monitor the deflection of points located in that horizontal plane and away from the bearing region. Each load test was repeated several times to average the values, and check that both the displacement and the reference values were the same each time within an uncertainty limit value of 0.015 mm. Then, the control points (Figure 2) were measured at the flaps of the ribs as well as measured at the corners of the CF-part in that horizontal plane since it is the most compromised position for its farther (maximum) distance to the bearing region.

3.1. Setup with Two CF-Parts

For the setup with two CF-parts, a *finger like stick* was applied to input a vertical load up to a maximum of 13.5 N —a limited value to avoid compromising the integrity of the walls. It was also checked that when the load applied had a tiny component off the vertical, this one caused a deviation within the displacement uncertainty.

The displacement was measured at two points at the horizontal plane (Figure 2a), and found a linear dependency on the load values within the uncertainty. The response of the models reproduced that trend, thus only the maximum displacement was of interest. It was also verified that deflection in other directions different than the vertical one did not appreciably disturbed either the setup or the models; therefore, only the vertical direction was monitored at the two points, one at the corner of the CF-part and the other one at the corner of the rib.

Furthermore, due to the weight caused by filling the CF-parts with the prismatic sensors, a *realistic loading* of the structure was defined. The material used had a rather high density of 4.5 g/cm^3, and the resulting weight was up to 2920 g per CF-part with an average value of 2645 g in comparison to the weight of the CF-part which was between 23 and 28 g. To describe this actual load situation, a second measurement was done by introducing steel blocks with a shape, weight, and centre of mass equal to the actual sensors inside the CF-parts.

3.2. Setup with Eight CF-Parts

For the same tile, the assembly with eight CF-parts was prepared (Figure 2b); this assembly resulted rather rigid and compact. It was not possible to load it with blocks without dismounting all parts, and, therefore, only point-like loads were applied. To produce the deflection, a load up to 40 N (again limited by the wall integrity) was applied at two different locations (green triangles in Figure 2b). The four points monitored with the control gauges at the horizontal plane were placed with two at the corners of the CF-parts and the other two at the corners of the rib; and this can be appreciated in Figure 2b, where the right front control gauge at the corner has been removed for a better view.

4. Definition and Analysis of the FE Models

All numerical calculations of FE models, as well as pre- and post-processing, were done with the software ANSYS (v15, ANSYS Inc., Canonsburg, PA 15317, USA), and using the composite pre-processing package (*ACP*). This package allows defining the CFRP layup and also couples the CFRP properties with those of the rest of materials within the solver. The ANSYS software is a widely deployed and well established tool for mechanical analysis including CFRP materials [12,16–18]. The orthotropic character of the CF-parts and the frictional contacts resulted in a nonlinear model which was solved by using the Newton–Raphson methods.

The FE models require defining the material properties, mesh, contacts, loads, and support conditions described in the following sections.

4.1. Material Properties

The walls of the CF-parts were made with two layers of epoxy-CFRP prepreg 1 K plain weave fabric of 0.15 mm thickness, with resin content of 40% (weight) produced by Cytec Industries Inc., (Woodland Park, NJ 07424, USA). The layers were set with different relative orientations, varying from face to face due to the wrapping of the fabric around the mould. The material properties depend on the fibre and matrix components, yarn size, 2D knitting of the yarns, and matrix impregnation. Manufacturers usually offer measured data for fabrics which are largely deployed in the industry. Therefore, final users have to perform their own tests according to appropriate standards [19–23] for other fabrics. This task is very time consuming and expensive.

However, micro-mechanical models [24] can provide estimates of the properties of the fabrics, sometimes with excellent predictive capabilities [25]. Authors in [26,27] developed a model for plain fabrics including the properties of the out-of-plane direction. The mechanical parameters defined according to the model described in [27] for the chosen fabric are listed in Table 1. Besides the material properties for the CFRP pre-processing, it was also necessary to introduce the fabric woven type, thickness, layers, and the orientation of its fibres in each layer of each face of the parts. The model of the CF-structure was made of 512 parts with five faces and two layers each. Therefore, this method can be used for small models like the two CF-parts setup, but, for larger models, the characterization of the orthotropic materials with such pre-processors would be largely time consuming.

Table 1. Mechanical properties of the CFRP fabric as determined according to the micromechanical model in [27]: Young's modulus (E), shear modulus (G) and Poisson's ratio (v). x and y-axis directions are in-plane. z-axis direction is out-of-plane (layer thickness).

Density [g/cm^3]	E_x,E_y [GPa]	E_z [GPa]	G_{xy} [GPa]	G_{yz},G_{xz} [GPa]	v_{xy}	v_{yz},v_{xz}
1.48	57.98	12.06	4.18	3.81	0.05	0.44

Orthotropic materials may be useful to counteract asymmetrical load patterns. In the case of this instrument, the load demand is isotropic, the CF-parts being located at any orientation in the structure, and the sensor weight acting as load. The selection of the fabric and layering design of CFRP parts was done to optimize the performance of the CF-part under the actual tri-axial loads. The double layer of plain fabric was selected to effectively achieve an isotropic behaviour [28]. Based on that assumption, the possibility to define an ad hoc isotropic material for the definition of the CFRP actual behaviour was studied. The properties of this *virtual* material were defined with the same Young's modulus value, and the shear modulus was adapted to reproduce the measured results. The value of the Poisson's ratio, linked to the previous parameters, resulted in a generic and reasonable value, causing no extra side-effects on the results. The values of the properties for this virtual material are listed in Table 2 as CFRP*.

Typical values found in literature for the metallic parts [29], ribs (steel AISI-304) and tiles (aluminium 5083) are also listed in the Table 2.

Table 2. Mechanical properties of the isotropic materials used in the models.

	Density [g/cm^3]	Young's Modulus [GPa]	Shear Modulus [GPa]	Poisson's Ratio
Aluminium 5083	2.67	71.0	26.69	0.33
Steel AISI 304	8.0	193.0	74.81	0.29
CFRP*	1.48	57.98	22.30	0.30

4.2. Mesh Model

The models included a finer mesh for the CF-parts because of its importance for the structure performance, and a higher-sized sized mesh for the metallic parts consistent with the very different deformations and rigidities of the parts. Typical mesh smoothing tools, as corrections for mapped surfaces, were used when possible.

The mesh of the tiles was done with 10-node tetrahedral elements. The element size was set to 15 mm with at least three nodes in its thinner zones, resulting in about 14,000 elements per part.

Since the ribs geometry is more irregular, a smaller size mesh of 5 mm made of hexahedral elements was used, resulting in about 21,000 elements per part.

For the CF-parts and due to its thin-wall geometry, it is recommended to use shell-elements for better efficiency and adaptation. However, the pre-processing package ANSYS-*ACP* requires solid elements, and therefore 8-node hexahedral elements of 1 mm in size were initially used, resulting in about 80,000 elements per part. A mesh-sensitive analysis was conducted to determine the adequate size of the elements for these parts also used in other models. The obtained results for a reference model of the setup with element sizes ranging from 1 to 8 mm showed deformation values clearly converging with decreasing element size. The relative differences were below 4.5% in the size ranging between 1 and 4 mm. However, the CPU calculation time increased by a factor seven in that same range. Because of this behaviour, the size value was set to 4 mm as a compromise for accuracy and CPU time, adding uncertainty to the results while assuming a deformation difference below 5%.

For models with (virtual) isotropic CFRP* material, a mesh of 20-node hexahedral elements was used, resulting in 17,350 elements per part.

In Figure 3, there are views of the mesh model used for the setup with two CF-parts. Figure 3a shows the mesh model of the assembly and Figure 3b shows a detail view of the region where two flaps of two ribs are grabbing the CF-part wall.

4.3. Model Boundary Conditions: Loads, Support and Contacts

In this subsection, the definitions for the loads, fixed support and contacts of the FE models are described.

In the mechanical test setup, the tile was bolted to a bearing support. Therefore, at the rib on the horizontal plane, a fixed strip-like region with width of 20 mm was defined for isostatic balance (Figures 4 and 5). The high rigidity of that region made no difference with respect to the actual mounting with a bolted joint of the tile, ribs, and bearing support (Figures 2, 4 and 5).

The load applied with a *finger-like stick* in the mechanical test setup became a point-like load in the model. For all models, these loads were applied on a reduced diameter region defined in the model; the value, direction, and location had to be as close as possible to the test conditions (Figures 4 and 5).

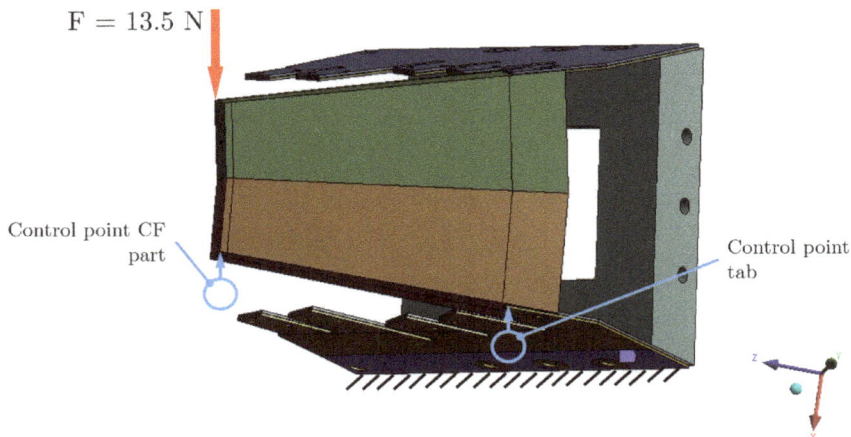

Figure 4. Drawing of the FE model used for the setup with two CF-parts. The point-like loads, fixed support and control points are marked.

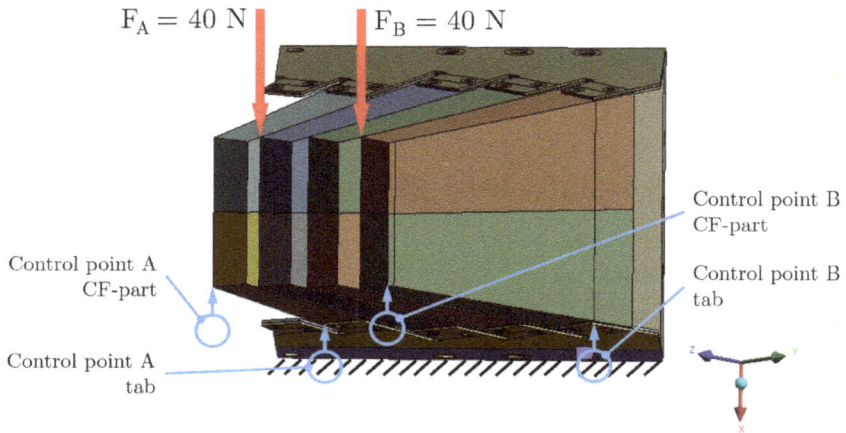

Figure 5. Drawing of the model used for the setup with eight CF-parts.

In the real structure, the actual sensors fill the CF-parts and cause the loading of the whole structure. The different orientations of the sensors cause the loading to lean over any of the CF-part faces depending on the region of the structure. The actual loading of the structure was a challenging problem for the FE model definition because the hundreds of new parts added. The model cannot use approximations such as *remote loads* due to the many orientations of the CF-parts, neither point-like masses instead of sensors due to different position of the centre of mass of CF-part and sensor. Therefore, the sensors and their weight were included in the models.

To study this problem, a real loading test was done cf. Section 3. The *realistic loading* for the two CF-parts setup was defined, where blocks to replace the sensors were added; thus, for the FE model, these blocks and their weight were included.

The study of the contacts [30] was another key for the adaptation of the models. Changes were implemented to relax the contact definitions while keeping the performance of the calculations within known limits as they were provided directly by the reference data.

Three contact regions were first studied: *RIB-RIB*, *RIB-TILE*, and *RIB-CFRP* (Figure 6). These parts were fastened with bolts; therefore, all these regions contain some bolted sub-region. The pressure-cone method was used with a cone-angle value of 45 degrees [31,32] to define the effective fastening region (bonded contact, linear type) which provided the right rigidity of the joint. Beyond the pressure-cone sub-region, models were studied with two types of contact: frictional (nonlinear) contact and with no-separation restriction (linear) contact allowing slipping. All contacts used a penalty-based formulation. Furthermore, frictional contacts follows Coulomb's friction law in order to define tangential stresses. Special attention was paid for nonlinear models when defining *master* and *slave* surfaces. In that case, the criterion was to define the stiffer surface as the master one, and therefore the CFRP material was defined as slave when in contact with metal parts (rib).

The *RIB-RIB* and *RIB-TILE* regions showed equivalent results with both contact types; this behaviour was most likely caused by the high rigidity of those regions, including only the metal parts. The contact type selected for these regions was the (linear) no-separation restriction. Then, to further study the possible linearization of the contacts, the focus was set to analyze the *RIB-CFRP* region, which was also fastened with bolts at the flaps. Different models were defined with linear and nonlinear contact types to analyse the different effects. The coefficient of friction between CFRP and the steel was set according to [33].

For the two CF-parts model, two load cases were studied, one with *point-like loads* and other with *realistic loads*. In the second case, a metal block was inserted inside each CF-part, which generated new contact regions. In the real case, even though the CF-structure held the sensors inside, these ones must be free and take no loads due to their fragility, and also to avoid compromising their integrity. In the model, to avoid these stiffening effects in the structure when adding blocks, the sensor mounting conditions were emulated by setting the material rigidity (Young's modulus) of the block to a negligible value. The contact *CFRP-BLOCK* was defined as a bonded (linear) type. The interplay of the definition of this contact and the rigidity is discussed in Section 5.1.

The CF-parts were glued together with epoxy resin and, to avoid including a complex glue-like contact with extra glue material between the CFRP walls, a bonded contact was used instead at the *CFRP-CFRP* region (Figure 6). However, for the model with eight CF-parts, this contact resulted in an extremely stiff structure. To better describe the wall gluing effect, the formulation of the bonded contact was done with control parameters for tangential and normal contact stiffness. This option is a possible approach to tackle contact problems [34], and it is available with ANSYS. Both tangential (FKT) and normal (FKN) control factors of this formulation can be set to values ranging from 0.01 to 1.0. Since the factors only tuned the bonded nature, the contact itself was kept. This procedure allowed for smoothing the bonded character of the contact, and the FKT and FKN values were set according to the results obtained in the models as compared with the test values. The factors of the contact must be common for both test cases, with two and eight CF-parts; however, and due to the rather small glued surface in the two CF-parts case, the gluing effect is limited. The minimum recommended value of 0.01 was used for FKN, and caused no material penetration issues in any setup. For the FKT values, the range from 0.1 to 1.0 caused no appreciable differences when applied to the models for two CF-parts, and, therefore, the value of 0.1 for FKT was set for all models.

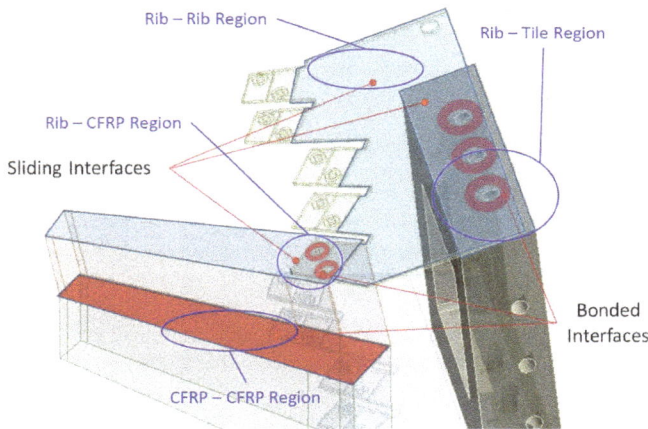

Figure 6. Drawing showing the contact regions at the two CF-parts set up. The pressure cone surfaces are marked, and contact regions indicated.

5. Results and Discussion for Mechanical Test Setups and FE Models

After defining the mechanical test setups and the configuration of the FE models, the results of the models were analysed and compared to the reference data to find a balance of a robust and simplified model still capable of providing realistic values of deformation and within a known uncertainty.

It is important to note that, in this work, only deformation values are presented and discussed. The reason for focusing on deformation is based on the immediate influence on the sensor position and the ultimate sensitivity of the calorimeter. From the mechanical design point of view, the stress is also a key parameter that has to be evaluated and considered to make any design decision, being directly related to safety concerns. A similar discussion as the one presented can be made to assess the stress conditions. However, it is beyond the scope of this work. It is only mentioned that the stress values obtained from the models were coherent with changes, and always far below any stress limits for safety.

5.1. Models with Two CF-Parts

In Figure 4, the drawing of the two CF-parts model is shown. The bearing region is defined as fixed at the horizontal (YZ) plane; the applied point-like load marked with a vertical red arrow is located at the upper corner of the CF-part, and the two control points in blue are positioned at the corners of the CF-part and the rib.

5.1.1. FE Model-1: CFRP Pre-Processing and Nonlinear Contacts

The reference model for this setup was defined with the pre-processing of the orthotropic material for the CF-parts and the nonlinear type contacts As explained in Section 4.3, for the *RIB-RIB* and *RIB-TILE* regions, the selection of no-separation (linear) or frictional (nonlinear) contact types caused no differences at this point. However, this model used a realistic frictional contact type at the *CFRP-RIB* region beyond the pressure cone of the bolts.

The values obtained in both control points resulted in being very similar to those obtained in the mechanical test setup, cf. *Model*-1 in Table 3, corroborating that both the FE model definition (including mesh and contacts) as well as the CFRP material characterization were highly reliable and provided robust results for further discussions.

Table 3. Measured and model results for the setup with two CF-parts and a point-like load.

	CFRP Pre-Process	Contact CFRP-RIB	Deflection Rib-Flap [mm]	Deflection CF-Corner [mm]
test			0.065	0.140
model -1	yes	nonlinear	0.069	0.142
model -2	no	nonlinear	0.119	0.151
model -3	no	linear	0.066	0.110

5.1.2. FE Model-2: CFRP* Isotropic Material and Nonlinear Contacts

Due to the high modelling time for the CFRP pre-processing, the possibility to re-define the CFRP as a (virtual) isotropic material was studied, considering that the fabrics and layers used would result in three-axial symmetric properties. As cited in Section 4.1, the CFRP material was redefined as a (virtual) isotropic CFRP* (cf. Table 2). The same set of contacts as in *Model*-1 was used, and the results showed a fair agreement at the CFRP corner control point (8% deviation relative to the measured value) and within the uncertainty. However, the result at the rib control point was poor when using the new material, which reduced the rigidity at that flap region.

5.1.3. FE Model-3: CFRP* Isotropic Material and Linear Contacts

The next step of the study was focused on the contact type at the *CFRP-RIB* region and, to simplify the calculation time, the contact was defined as a no-separation (linear) type. This modification is expected to induce a higher rigidity in the structure and counterbalance the low rigidity of *Model*-2 due to the material re-definition. Additionally, the new model was the most simplified due to the full linearization of all the contact types, including the isotropic character of the virtual material CFRP*.

The results at the CFRP corner control point differed by 21%, and at the flap control point resulted in being very similar to the measured value. Therefore, the deviations were proportional to the distance between the control points and the bearing region. As expected, the trade-off of effects of the combined approximations for material and contact definitions resulted in being effective since the functional description of the FE model was kept even though it was rather simplified.

5.1.4. FE Model-4: CFRP* Isotropic Material and Realistic Loads

By comparing the results, it can be stated that the FE *Model*-3, including the properties of the (virtual) isotropic CFRP* material, as well as linear-type contacts between the parts, provides reliable results of deformations within the uncertainty limit of 21% at the point of maximum deflection of the CF-part located away from the bearing regions.

Based on *Model*-3, a new FE model was studied, where the type of applied loads was changed to the weight caused by sensor-like blocks. As previously mentioned in Section 4.3, when changing to realistic loads, a new contact region arose. This new contact *CFRP-BLOCK* was defined as a bonded (linear) contact type. The use of this contact for the CFRP-block region raised the question of a possible deformation of the CF-part walls because of the combination of a low-rigidity value assigned to the block (cf. Section 4.3) and the thin-walls of the CF-parts. An independent evaluation of the *CFRP-BLOCK* contact was defined as a realistic frictional contact and the actual rigidity of the material provided wall deformation values rather similar (below 7% deviation) to the selection presented here. More importantly, the iso-surfaces of the stress and deformation distributions showed the same pattern and had comparable values for both models. Therefore, the linear option was selected as the most adequate for our purposes.

The results provided by this model are listed in Table 4. The displacement results are similar to the obtained in the mechanical test setup within 6% deviation at the rib control point and 9% deviation at the CFRP corner control point.

The use of distributed loads resulted in a better performance of the model. Even though the reference *Model*-1 under point-like load resulted in being very accurate, the CFRP pre-processing causes limitations to deploy to many parts' structures. On the other hand, *Model*-3 with its distributed (realistic) load counterbalanced the limitations caused by modifying the contact character, and therefore it was more adequate for the description of the whole structure.

Table 4. Data and results of the model for the setup with two CF-parts and sensor-like blocks.

	CFRP Pre-Process	Contact CFRP-RIB	Deflection Rib-Flap [mm]	Deflection CF-Corner [mm]
test			0.266	0.558
model-4	no	linear	0.283	0.505

5.2. Model with Eight CF-Parts

The eight CF-parts model allowed us to perform the study of the effects produced by increasing size and number of parts in the FE model, as well as the evaluation of these scaling effects for the whole structure. For the eight CF-parts model, a point-like load was applied (Figures 2b and 5). The bearing region located at the horizontal plane (YZ) was defined as fixed. At the edge between two parts and away from the bearing region, point-like loads (red vertical arrow) were applied at two different positions (*A* and *B*) for independent tests with four control points (marked in blue) at the corners of the CF-parts and the rib located at the horizontal plane.

The FE model for this setup was done according the *Model*-3 definitions for the 2 CF-parts setup. The most important differences in this model were the large regions with *CFRP-CFRP* contact caused by the gluing of all the CF-parts, and defined as discussed in Section 4.3. The deformation values were calculated and compared with the measured data for the two independent load test at points *A* and *B* (see Table 5). The values differed at the rib point ranging from 0 to 21% (relative values), and for the CFRP corners ranging from 48 to 53%. The models always provided lower deformation values corresponding to a more rigid structure than the real one. This stiffening effect was observed in *Model*-3 (cf. Table 3), with a deviation of 21%, as well as in the model with the realistic load (Table 4), which had a deviation of 9%, both referring to the control point at the CF corner. Therefore, this stiffening effect was mostly due to the material and contact selected, and was in no case caused by the large glue-contact. Setting the *CFRP-CFRP* contact parameters (FKT and FKN) to negligible values did not notably change the results and, consequently, demonstrated the robustness of the model to this tuning selection.

Looking only at the load point at side *B*, the results of the setups with two and eight CF-parts (Table 3 and Table 5) can be compared. The deformation, within the uncertainty, was similar in both cases even though, in the second case, the load applied was three times higher. Clearly and as expected, when more parts were added to form the structure, the stiffer it became. Moreover, when the load was applied at side *A*, the deformation values obtained were similar to those in the case where the load was applied at side *B* for opposite points; therefore, these results showed that both the model and the setup were highly symmetrical despite the real geometry.

Table 5. Data and results of the model for the setup with eight CF-parts and point-like loads.

	Load Point	Pre-Process	Contact CFRP-RIB	Deflection Rib-Flap		Deflection CF-Corner	
				A [mm]	B [mm]	A [mm]	B [mm]
test	A			0.070	0.040	0.160	0.100
model	A	no	linear	0.058	0.040	0.076	0.052
test	B			0.050	0.070	0.120	0.130
model	B	no	linear	0.043	0.055	0.055	0.066

6. FE Model for the Instrument Structure

After studying all the selections applied to the FE models previously mentioned in the sections above, the main objective of this research can be tackled by evaluating an FE model for the calorimeter structure as a whole, considering just one half of the structure due to its symmetry at the vertical mid-plane (Figures 1d and 7).

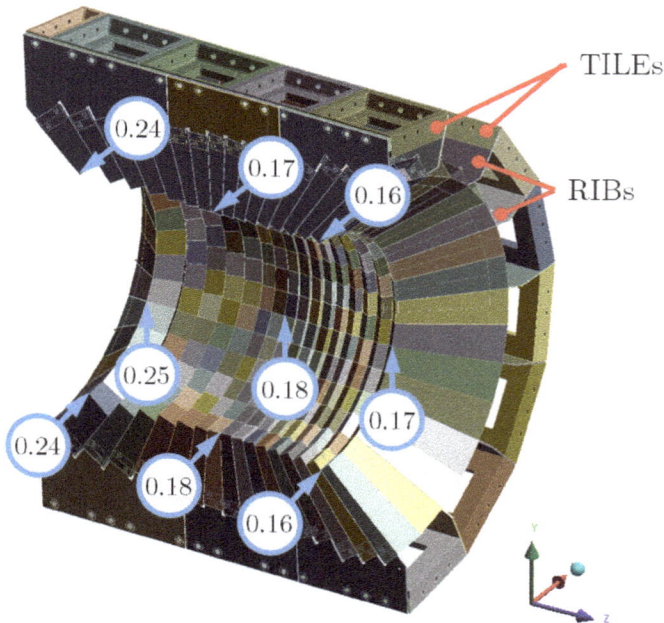

Figure 7. Drawing of the structure showing the maximum nodal displacements at the corners of the CF-structure.

The material was defined as CFRP*; the contacts were defined as bonded type at the pressure cones for the bolted joints, tuned bonded types for the CFRP–CFRP regions and non-separating type in any other region; the sensors weight were defined as the applied load. The support of the system was also defined as if the cover was fixed at some faces to an external frame [14]. The results of the model calculations showed a negligible deformation (well below 0.01 mm) of the stiff cover structure parts. Therefore, for the evaluation of the structure design, the points with the maximum deflection were selected, which were the nodal points located at the corners of the CF-parts in the internal opening and away from the bearing region. In Figure 7), the values calculated for different corners of the CF-structure are shown. The points correspond to the upper and bottom positions at the mid-vertical plane and the mid-horizontal plane.

Then, taking into account the maximum deformations in the CF-structure, the reliability, modelling, and computing time for the whole structure were determined.

6.1. Model Deformations

The deformation values at the nodal points were rather similar in each half-ring, the variation being just a few *micrometers* along the same half-ring. The values were slightly higher at the lower regions with respect to the upper one. This difference was caused by the cumulative effect of the weight acting from top-to-bottom at each half-ring. The tiny differences show the capability of the honeycomb structure to provide stiffness in every region of the CF-structure.

The deformation values also changed along the longitudinal axis from 0.16 up to 0.24 mm (Figure 7). The different position of the rib-flaps relative to the bearing region at the tiles changed the lever arm and caused the different deformation along the axis.

The tri-axial deformation at the CF corner points determined in which direction the structure was deflected and which region was affected. Along each half-ring, the deformation changes direction from top to bottom: at the upper region, the deflection was mostly horizontal and in the positive x-axis direction; at the mid-horizontal plane, the deformation was in the vertical direction and downwards (negative y-axis); and, at the lower region, the deflection was again horizontal but pointing to the negative x-axis direction. These directions are expected for the loading caused by the weight. This effect combined to the axial difference, which makes a torsional-like deformation in the CF-structure. The high rigidity of the CFRP structure made the deformation very limited in value at the most unfavorable locations.

6.2. Model Reliability

For the model calculations, the comparison with the measured data defined the uncertainty, ranging from 9 to 53% for the most critical points, the CFRP corners far away from the bearing regions. All these depended on the load cases, where the lower value corresponded to the realistic load distribution, and also depended on the type of structure with either two or eight CF-parts.

Cumulative uncertainties for the whole structure were tough to trace back. On the one hand, the stiffening effect inherited in the model grew faster in the model than in the mechanical setup, with the increasing number of parts. It showed deviations up to 21% and 53% for two and eight CF-parts, respectively (cf. Tables 3 and 5). On the other hand, the use of a realistic load distribution resulted in a difference reduced by a factor 2.3 (21% to 9% deviation, cf. Tables 3 and 4). This effect was not expected to be related to the structure size, but to the load nature. Due to the addition of parts and the application of real load distributions, and based on the previous discussion, a sound deviation up to 370% was estimated as possible. This evaluation gives a rough estimated maximum deformation value of 0.9 mm at the CFRP corner points.

To have a better picture of the quality of these results, the uncertainty linked to the production of the parts, the bundles of parts and the mounting of the whole structure has to be considered [15]. The inner volume of the parts was obtained with tolerances of 0.05 mm in any plane; however, the outer surfaces and edges were not so tightly constrained. Additionally, the CF-parts were glued together in a full handcraft process. The measured fluctuations in envelope dimensions of eight CF-parts reached up to 1 mm, and similar values are typical for mounting the parts into the cover structure. According to the FE model and assuming maximal uncertainties, the expected deformation values (0.9 mm) remained well below the added tolerance values of production and assembly added together (2 mm). This situation showed that the model evaluation was properly suited to provide design decisions, even considering the largest uncertainties.

6.3. Modelling and Computing Time

One of the key goals for the development of this model was the possibility to have an acceptable modelling and computing time. Complex mechanical calculations tough to define and containing thousands of features are hardly useful to explore design options in an effective way during the timeline of typical projects. In our model, the use of a material pre-processor (a critical model definition step) was avoided, and only linear contact sets were used, guaranteeing the convergence despite the large number of features.

The time involved in the pre- and post-processing stages of the model was evaluated to be between 20 and 25 h, using ANSYS or similar software packages, which typically included *boosting* tools, e.g., to define symmetrically located contacts. The calculation took 38.3 h based on a single CPU *Intel Core* i7-4720HQ (Intel Corporation, Santa Clara, CA 95054-1549, USA), 3.60 GHz, 16 Gb RAM, and ANSYS v15 (64 bits). The total time is fully compatible with a 10-day time frame dedicated to mechanical calculations, including the model definition, revision, evaluation, and analysis.

7. Conclusions

This work analysed a case study involving initially a highly nonlinear FE model for a complex thin-wall structure made of CFRP materials with the goal to make an FE model robust and reliable enough as to be used as a design tool for the study stage of one scientific instrument. The complexity of the FE model arising from the many parts, features, and details was a tough bottleneck for implementing such FE model to work in a flexible and effective way and provide the right feedback to the rest of the project stages.

For the validation of the FE model, two mechanical setups, one of two CF-parts and the other with eight CF-parts, were mounted. The FE models studied allowed for selecting the right set of contact definitions and the CFRP properties as (virtual) isotropic material, so that the displacement results obtained were within 9% deviation for a realistic load.

With these considerations, a full linear model was successfully defined with sound assumptions, capable of reproducing the results within known deviations, and, once extended to the whole structure, yielded the next results:

- The overall time necessary for the model definition and calculation fits in a reasonable time span to interact with the project needs, despite the large amount of parts and contacts defined in the FE model.
- The total deformation calculated provided values below 0.24 mm in the key regions. The overall uncertainty expected from small setups to the whole structure was estimated and a sound maximal deformation was set (0.9 mm). For comparison, the production and mounting tolerance values remain much higher (220% factor). Therefore, the FE model can definitely provide useful results for the design project.

The kind of questions tackled down with this FE model were design proposals raised through the project process, including large volume changes of CF-parts, CFRP wall thickness limits, CF-part shapes, etc. [15]. The methodology developed can be deployed to other cases demanding intensive mechanical evaluations of complex (many parts) structures including CFRP materials.

Author Contributions: Conceptualization, E.C. and J.A.V.; Methodology, E.C. and A.S.; Software, J.C.R. and J.A.L.-C.; Validation, E.C. and A.S.; Investigation, E.C. and J.C.R.; Writing—Original Draft Preparation, E.C.; Writing—Review and Editing, E.C. and A.S.; Visualization, J.C.R.; Supervision, E.C.; Project Administration, E.C.; Funding Acquisition, E.C. and J.A.V.

Funding: This work was partially supported by the *Ministerio de Economía y Competitividad* (Spain), reference FPA2015-69640-C2-2-P, and by *Xunta de Galicia* (Spain) references EM2012/140 and GRC2015/016.

Acknowledgments: The authors thank A. Iglesias-Moreira who defined the basic ACP 2-parts model in her Bachelor Thesis (University of Vigo, Spain, 2014).

Conflicts of Interest: The authors declare no conflict of interest.

Abbreviations

The following abbreviations are used in this manuscript:

FE	finite element
CFRP	carbon-fibre reinforced plastic
CF	carbon fiber
CFRP*	(virtual) isotropic material
FKT	bonded contact stiffness factor, tangential component
FKN	bonded contact stiffness factor, normal component
RIB-TILE	contact region of the rib and tile parts
RIB-RIB	contact region of two rib parts
RIB-CFRP	contact region of the rib and CFRP parts
CFRP-BLOCK	contact region of the CFRP and block parts
CFRP-CFRP	contact region of two CFRP parts

References

1. Kassapoglou, K. *Design and Analysis of Composite Structures*, 2nd ed.; Wiley & Sons: London, UK, 2013; ISBN 978-1-118-40160-6.
2. Breuer, U.P. *Commercial Aircraft Composite Technology*; Springer: Cham, Switzerland, 2016; ISBN 978-3-319-31917-9.
3. Walker, M. Optimal Design of Fibre-reinforced Laminated Plates Accounting for Manufacturing Uncertainty. *Int. J. Mech. Mater. Des.* **2005**, *2*, 147–155. [CrossRef]
4. Yin, H.; Wen, G. Theoretical prediction and numerical simulation of honeycomb structures with various cell specifications under axial loading. *Int. J. Mech. Mater. Des.* **2011**, *7*, 253–263. [CrossRef]
5. El-Abbasi, N.; Meguid, S.A. A Continuum Based Thick Shell Element for Large Deformation Analysis of Layered Composites. *Int. J. Mech. Mater. Des.* **2005**, *2*, 99–115. [CrossRef]
6. Jessen, N.C.; Norgaard-Nielsen, H.U.; Schroll, J. CFRP lightweight structures for extremely large telescopes. *Compos. Struct.* **2008**, *82*, 310–316. [CrossRef]
7. Hick, M.T. *Design of a Carbon Fiber Composite Grid Structure for the GLAST Spacecraft*; SLAC Report 575; Stanford University: Stanford, CA, USA, 2001.
8. Morozov, E.V.; Lopatin, A.V. Analysis and design of the flexible composite membrane stretched on the spacecraft solar array frame. *Compos. Struct.* **2012**, *94*, 3106–3114. [CrossRef]
9. Olcesea, M.; Caso, C.; Castiglioni, G.; Cereseto, R.; Cuneo, S.; Dameri, M.; Claudia, G.; Glitza, K.-W.; Lenzen, G.; Mora, M.; et al. Ultra-light and stable composite structure to support and cool the ATLAS pixel detector barrel electronics modules. *Nucl. Instrum. Meth. A* **2004**, *518*, 728–737. [CrossRef]
10. Giraudo, G.; Borotto, F.; van den Brink, A.; Buskop, J.; Coli, S.; Feofilov, G.; Igolkine, S.; Nooren, G.J.L.; Riccati, L. The SDD and SSD support structure for the ALICE Inner Tracking System. *J. Instrum.* **2009**, *4*, P01003. [CrossRef]
11. Wieman, H.H.; Anderssen, E.; Greiner, L.; Matis, H.S.; Ritter, H.G.; Sun, X.; Szelezniak, M. STAR PIXEL detector mechanical design. *J. Instrum.* **2009**, *4*, P05015. [CrossRef]
12. de Lorenzis, L.; Grancagnolo, F.; L'Erario, A.; Maffezzoli, A.; Miccoli, A.; Rella, S.; Spedicato, M.; Zavarise, G. Analysis and characterization of the mechanical structure for the i-tracker of the mu2e experiment. *Nucl. Phys. B* **2014**, *248–250*, 134–136. [CrossRef]

13. Aumann, T. Technical Report for the Design, Construction and Commissioning of The CALIFA Barrel. 2011. Available online: https://edms.cern.ch/document/1833500/1 (accessed on 4 February 2019).

14. Casarejos, E.; Alvarez Pol, H.; Cortina-Gil, D.; Durán, I.; Izquierdo, P.; Yáñez Alfonso, P.; Vilan, J.A. The mechanical design of the BARREL section of the detector CALIFA for R3B-FAIR. *EPJ Web Conf.* **2014**, *66*, 11037. [CrossRef]

15. Casarejos, E.; Alvarez Pol, H.; Cortina-Gil, D.; Durán, I.; Izquierdo, P.; Yáñez Alfonso, , P.; Vilan, J.A. Design and construction of the structure of the DEMONSTRATOR of the CALIFA detector for R3B-FAIR using carbon-fiber composites. *EPJ Web Conf.* **2014**, *66*, 11038. [CrossRef]

16. Meng, M.; Le, H.R.; Rizvi, M.J.; Grove, S.M. The effects of unequal compressive/tensile moduli of composites. *Compos. Struct.* **2015**, *126*, 207–215. [CrossRef]

17. Wen, Y.; Yue, X.; Hunt, J.H.; Shi, J. Feasibility analysis of composite fuselage shape control via finite element. *J. Manuf. Syst.* **2018**, *46*, 272–281. [CrossRef]

18. Komarov, V.A.; Kurkin, E.I.; Spirina, M.O. Composite aerospace structures shape distortion during all stages of vacuum infusion production. *Procedia Eng.* **2017**, *185*, 139. [CrossRef]

19. ASTM International. *Standard Test Method for Tensile Properties of Plastics (D638)*; ASTM International: West Conshohocken, PA, USA, 2015.

20. ASTM International. *Standard Test Method for Compressive Properties of Polymer Matrix Composite Materials with Unsupported Gage Section by Shear Loading (D3410)*; ASTM International: West Conshohocken, PA, USA, 2016.

21. ASTM International. *Standard Test Method for Shear Properties of Composite Materials by the V-Notched Beam Method (D5379)*; ASTM International: West Conshohocken, PA, USA, 2012.

22. ASTM International. *Standard Test Methods for Flexural Properties of Unreinforced and Reinforced Plastics and Electrical Insulating Materials (D790)*; ASTM International: West Conshohocken, PA, USA, 2017.

23. ASTM International. *Standard Test Method for Tension-Tension Fatigue of Polymer Matrix Composite Materials (D3479)*; ASTM International: West Conshohocken, PA, USA, 2012.

24. Dixit, A.; Harlal Singh, M. Modeling techniques for predicting the mechanical properties of woven-fabric515 textile composites: A Review. *Mech. Compos. Mater.* **2013**, *49*, 1–20. [CrossRef]

25. Ryou, H.; Chung, K.; Yu, W.-R. Constitutive modeling of woven composites considering asymmetric/anisotropic, rate dependent, and nonlinear behavior. *Compos. Part A* **2007**, *38*, 2500–2510. [CrossRef]

26. Naik, N.K.; Ganesh, V.K. An analytical method for plain weave fabric composites. *Composites* **1995**, *26*, 281–289. [CrossRef]

27. Chretien, N. Numerical Constitutive Models of Woven and Braided Textile Structural Composites. Master's Thesis, Faculty of Virginia Polytechnic Institute and State University, Blacksburg, VA, USA, 2002.

28. Hosoi, A.; Kawada, H. Fatigue Life Prediction for Transverse Crack Initiation of CFRP Cross-Ply and Quasi-Isotropic Laminates. *Materials* **2018**, *11*, 1182. [CrossRef]

29. *MatWeb Database*; MatWeb LLC: Blacksburg, VA, USA, 2018.

30. Meguid, S.A.; Czekanski, A. Advances in computational contact mechanics. *Int. J. Mech. Mater. Des.* **2008**, *4*, 219–443. [CrossRef]

31. Ito, Y.; Toyoda, J.; Nagata, S. Interface pressure distribution in a bolted flange assembly. *J. Mech. Des.* **1979**, *101*, 330–337. [CrossRef]

32. Budynas, R.G.; Nisbett, J.K. *Shigley's Mechanical Engineering Design*, 5th ed.; McGraw-Hill: New York, NY, USA, 2015; ISBN 978-0073398204.

33. Cornelissena, B.; Warnet, L.; Akkerman, R. Friction measurements on carbon fibre tows. In Proceedings of the 14th International Conference on Experimental Mechanics (ICEM), Poitiers, France, 4–9 July 2010.

34. Zabulionis, D.; Rimsa, V. A Lattice Model for Elastic Particulate Composites. *Materials* **2018**, *11*, 1584. [CrossRef] [PubMed]

Article

Optimal Design of Sandwich Composite Cradle for Computed Tomography Instrument by Analyzing the Structural Performance and X-ray Transmission Rate

Sang Jin Lee [1,2] and Il Sup Chung [2,*]

1   Composites Convergence Team, Korea Textile Machinery Convergence Research Institute, Gyeongsan 38542, Korea; sjlee@kotmi.re.kr
2   Department of Mechanical Engineering, Yeungnam University, Gyeongsan 38541, Korea
*   Correspondence: ilchung@yu.ac.kr; Tel.: +82-53-810-3525

Received: 15 December 2018; Accepted: 14 January 2019; Published: 17 January 2019

Abstract: Carbon fiber-reinforced composite has an excellent X-ray transmission rate as well as specific modulus and strength. The major components of medical devices, X-ray systems, and computed tomography (CT) equipment that require superior X-ray transmission performance also require structural performance for deformation. Therefore, medical components consist of a sandwich composite structure with carbon fiber composites applied as a face material. The X-ray transmission ratios of face material and foam material were measured according to thickness, and the relation equation for thickness and X-ray transmission rate was derived. The X-ray transmission rate for the sandwich composite structure consisting of face and core material was measured and the relationship between the X-ray transmission rate and the dimension for thickness of sandwich cradle was derived. Using the optimization process, the thicknesses of face and core materials for sandwich cradles were determined to minimize the cost of used materials. They also met the criteria that the deflection should not be more than 20 mm, and the X-ray transmission rate of the cradle should be equal to or greater than that of aluminum at 1.5 mm thickness. The sequence pattern of face material was proposed through structural analysis. The face material of the sandwich cradle was manufactured by a resin infusion and vacuum bagging method, followed by inserting the core between the cured faces. Next, the sandwich cradle assembly was completed and re-cured. The sandwich cradle met the criteria that the deflection at the end was 19.04 mm and the X-ray transmission was 78.7% greater than the X-ray transmission of 1.5 mm aluminum.

Keywords: computed tomography; sandwich composite; X-ray transmission; CT cradle

1. Introduction

Historically, composite materials were lightweight and of superior stiffness and strength. Composite materials were used in aircraft parts and defense components where such characteristics are required [1]. In recent years, their applications have been widely extended to the structures of trains, buses, and other vehicles [2–7]. Computed Tomography (CT) is widely used in non-destructive inspections of laminate and sandwich composites [8–15]. Since the early 2000s, composites have been applied to medical device components and are now being used to satisfy X-ray transmission performance and weight requirements. The components of CT and diagnostic X-ray equipment requiring X-ray transmission performance are produced using carbon material. Carbon fiber reinforced plastics (CFRP) components and plates were designed by several optimization processes for the shape and thickness [16–20].

In this paper, the development process of the sandwich cradle for medical devices composed of carbon fiber reinforced plastics CFRP for the face material and foam material as the core was explored. The X-ray transmittance analysis was conducted experimentally to measure the transmission rate according to the thickness of the materials for face material and core material. The specifications were determined to meet the criteria for structural performance and to minimize the cost of applied material for the sandwich cradle. The face material was formed using an infusion and vacuum bagging method. Next, the foam core was inserted between the faces. Finally, the sandwich cradle was assembled and cured to ensure uniform X-ray transmission performance to the X-ray measurement area of the sandwich cradle.

As the purpose of this study, the designing technology of a CFRP sandwich cradle with better X-ray transmission performance than the conventional metallic cradle and the fabricating process of a CFRP sandwich cradle for the uniform X-ray transmission are examined.

2. Fundamental Properties of Materials

Among the used fiber materials, CU 125NS (HANKUK CARBON Co., Ltd, Miryang, Korea) and MCU 125NS (HANKUK CARBON Co., Ltd, Miryang, Korea) were unidirectional (UD) prepreg 0.153 mm and 0.155 mm thick, respectively, as shown in Table 1. Plain woven type CF 3327 (HANKUK CARBON Co., Ltd, Miryang, Korea) fabric had a warp density of 11 threads/inch and a weft density of 11 threads/inch. The thickness of CF 3327 is 0.25 mm. CU 125NS was used as the UD carbon fiber, MCU 125 NS was considered to have excellent mechanical properties, and plain-woven fabric CF 3327 was used as the facing material. The core material used was Polymethacrylimide (PMI) foam. Mechanical properties of the materials used are shown in Table 2.

Table 1. Specifications of applied fiber.

Material	Carbon Fiber Weight (g/m^2)	Resin Weight (g/m^2)	Resin Contents (%)	Thickness (mm)
CU 125NS Prepreg *	125	62	33	0.153
MCU 125NS Prepreg *	125	64	34	0.155
CF 3327 **	205	105	33.8	0.25

* The density of CU 125NS (Carbon Uni-directional, 125 g/m^2, No Scrim) composite cured by vacuum bagging is 1500 kg/m^3. The density of MCU 125NS (Medium modulus Carbon Uni-directional, 125 g/m^2, No-Scrim) is 1500 kg/m^3. ** CF 3327 (Carbon Fabric) composite was fabricated by infusion and the density was similar to 1500 kg/m^3.

Table 2. Mechanical properties of applied materials.

Property Material	Tensile Modulus (GPa)	Shear Modulus (GPa)	Poisson's Ratio	Strength (MPa)
CU 125NS	$E_1 = 127.6$ $E_2 = 7.58$	$G_{12} = 4.05$	$V_{12} = 0.34$ $V_{21} = 0.05$	2650 65.5
MCU 125NS	$E_1 = 191.4$ $E_2 = 8.58$	$G_{12} = 5.10$	$V_{12} = 0.24$ $V_{21} = 0.04$	-
CF 3327	$E_1 = 48.3$	$G_{12} = 3.81$	$V_{12} = 0.07$	548.9
PMI foam (31 IG)	0.036	0.013	0.02	1.0

3. X-ray Transmission of Materials

3.1. Configuration of Test Device

The X-ray transmission equipment used in performance evaluation is a digital X-ray system made by Listem Inc (Wonju-Si, Korea). The configuration of the device for performance evaluation of X-ray transmission is shown in Figure 1. The distance between the source and detector is 1000 mm and the test sample was placed in the center. The field size is 40 mm × 40 mm. The conditions of X-ray transmission were 100 kV, 200 mA, and an exposure time of 0.1 s.

Figure 1. Test configuration of X-ray transmission.

3.2. Transmission Performnace of Face materials

The face materials were chosen from 6 types of UD materials manufactured through vacuum bagging, 2 types of UD + fabric prepreg through vacuum bagging, and 4 types of fabric materials made by infusion as shown in Table 3. The transmission rate was measured as the initial value (I_0) without specimen applied for 0.1 s at 100 kV, 200 mA and the transmission value (I_1) measured through each specimen was analyzed under the same conditions. The transmission rate is the ratio of I_1/I_0. In the case of the CU 125NS composite, the X-ray transmission rate decreased rapidly from 96.8% to 74.7% as thickness increased from 0.7 mm to 4.5 mm. Regardless of the type of molding technique and material, the transmission rate was approximately the same for the same thickness. Therefore, the relation between the transmission rate and thickness of face material is expressed in the following equation and predicted in Figure 2.

$$\text{Transmission rate of face material (\%)} = -5.7752 \times \text{thickness (mm)} + 100.89 \qquad (1)$$

Table 3. X-ray Transmission ratio of CFRP.

	Fabrication	Material of Face	Total Thickness (mm)	Measured Value (I_1) (unit: mRem)	Ratio of Transmission (I_1/I_0 *) × 100	Remark
Face material	Prepreg vacuum bagging	100% CU 125NS	0.75	214	96.8	A(1)
		100% CU 125NS	1.5	203	91.9	A(2)
		100% CU 125NS	2.25	194	87.8	A(3)
		100% CU 125NS	3	186	84.2	A(4)
		100% CU 125NS	3.75	175	79.2	-
		100% CU 125NS	4.5	165	74.7	-
	Prepreg vacuum bagging	33.3% CU 125NS + 66.7% CF 3327	2.25	194	87.8	B(1)
		60% CU 125NS + 40% CF 3327	3.75	175	79.2	B(2)
	Infusion	100% CF 3327 fabric	0.75	213	96.4	-
		100% CF 3327 fabric	1.5	202	91.4	C(1)
		100% CF 3327 fabric	3	185	83.7	C(2)
		100% CF 3327 fabric	4.5	165	74.7	C(3)

* The value of I_0 is 221. The densities of above composites were controlled as 1500 kg/m^3.

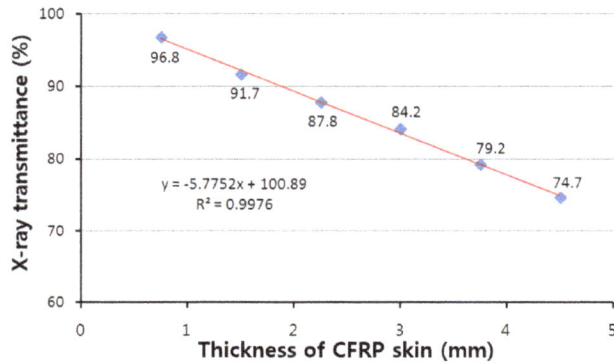

Figure 2. X-ray Transmission ratio of facing materials.

3.3. Transmission Performnace of Core Materials

The X-ray transmission rate of core materials was compared for different thicknesses and densities of foam. These were compared and analyzed for PMI foam 31IG with a density of 32 kg/m^3 and PVC (Polyvinyl chloride) foam with a density of 40 kg/m^3. Thicknesses of 2, 11, 30, and 45 mm were considered for PMI Foam 31 IG and an evaluation was performed for PVC foam at thicknesses of 30 and 50 mm. For the PMI 31IG foam, as the thickness of core increased, the transmission rate decreased slightly as shown in Table 4 and Figure 3. PVC foams with a density of 40 kg/m^3 showed a transmission rate of 90.1% for a thickness of 30 mm when compared to the transmission rate of PMI 31IG with the same thickness. The relationship between the transmission rate and thickness of core material for PMI 31IG is explained in the following equation and predicted in Figure 3.

$$\text{Transmission rate of core material}(\%) = -0.0822 \times \text{thickness (mm)} + 99.659 \tag{2}$$

Table 4. X-ray Transmission ratio of foam material.

	Material	Density (kg/m^3)	Total Thickness (mm)	Measured Value (unit: mRem)		Ratio of Transmission $(I_1/I_0) \times 100$
				I_0	I_1	
Core	PMI foam, 31 IG	32	2	221	220	99.5
			11	223	220	98.7
			30	223	217	97.3
			45	223	214	95.9
	PMI foam, 51 IG	51	45	223	208	93.3
			60	223	203	91.2
	PVC foam	50	45	223	209	93.7
			60	223	204	91.5

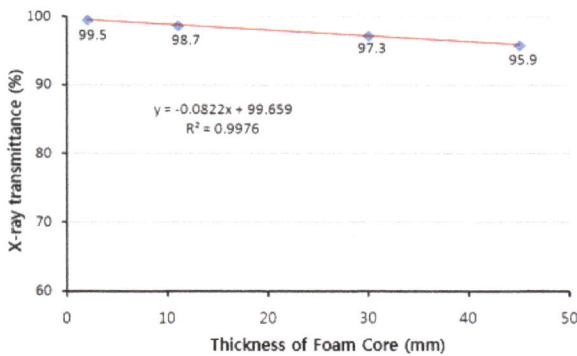

Figure 3. X-ray Transmission ratios of core material.

3.4. Transmission Performance of Sandwich Structures

Using the results for X-ray transmission performance of face materials and core materials, several cases of sandwich structure were constructed as shown in the Table 5 and the transmission rate of X-ray was assessed in each case on the same conditions from paragraph 3.1. Through this process, the correlation between the X-ray transmission rate of individual materials and the X-ray transmission rate of sandwich structures was derived. The relationship between the measured value of the actual sandwich structure and the transmission rate calculated using Equation (1) for total thickness of face and Equation (2) for the thickness of core was analyzed in Table 5. The X-ray transmission rate of sandwich structure is estimated by multiplying that of face material calculated by Equation (1) and that of core material calculated by Equation (2). Most cases agree with the following Equation (3) except for some structures made of PVC foam and a thick face.

$$\begin{aligned} \text{Transmission rate of sandwich structure} (\%) \\ = (\text{Transmission rate for total thickness of face calculated from Eq.}(1)) \\ \times (\text{Transmission rate for the thickness of core calculated from Eq.}(2)) \end{aligned} \quad (3)$$

Table 5. X-ray Transmission Ratios for Sandwich Structure.

Sandwich Structure * (face/core/face)	Total Thickness (mm)	Measured Value (unit: mRem)		Ratio of Transmission $(I_1/I_0) \times 100$	Transmission Rate of Face Calculated From Eq.(1)	Transmission Rate of Core Calculated From Eq.(2)
		I_0	I_1			
A(1) + PMI(31 IG, 11t) + A(1)	12.5	223	192	86.1	92.2	98.7
A(2) + PMI(31 IG, 11t) + A(2)	14	223	178	79.8	83.5	98.7
A(1) + PMI(51 IG, 45t) + A(1)	46.5	223	193	86.5	92.2	96.0
A(2) + PMI(51 IG, 45t) + A(2)	48	223	177	79.4	83.6	96.0
A(2) + PVC(30t) + A(2)	33	220	164	74.5	83.6	97.2
A(2) + PVC(50t) + A(2)	53	220	163	74.1	83.6	95.5
B(1) + PMI(51 IG, 45t) + B(1)	49.5	220	160	72.7	74.9	96.0
B(2) + PVC(30t) + B(2)	37.5	222	144	64.9	57.6	97.2
B(2) + PVC(50t) + B(2)	57.5	222	143	64.4	57.6	95.5
C(1) + PMI(31 IG, 11t) + C(1)	14	222	178	80.2	83.6	98.7
C(2) + PMI(31 IG, 11t) + C(2)	17	222	166	74.8	66.2	98.7
C(1) + PMI(51 IG, 45t) + C(1)	48	222	177	79.7	83.6	96.0
C(2) + PMI(51 IG, 45t) + C(2)	51	222	160	72.1	66.2	96.0
C(2) + PVC(30t) + C(2)	36	222	150	67.6	66.2	97.2
C(2) + PVC(50t) + C(2)	56	222	148	66.7	66.2	95.5

Table 5. *Cont.*

Sandwich Structure * (face/core/face)	Total Thickness (mm)	Measured Value (unit: mRem)		Ratio of Transmission $(I_1/I_0) \times 100$	Transmission Rate of Face Calculated From Eq.(1)	Transmission Rate of Core Calculated From Eq.(2)
		I_0	I_1			
C(1) + PMI(31 IG, 45t) + A(3)	48.75	222	172	77.5	79.2	96.0
C(1) + PVC(50t) + A(3)	53.75	222	159	71.6	79.2	95.5
C(2) + PMI(51 IG, 45t) + A(2)	49.5	222	168	75.7	74.9	96.0
C(2) + PVC(50t) + A(2)	54.5	222	156	70.3	74.9	95.5
B(1) + PMI(51 IG, 45t) + B(2)	51	222	160	72.1	66.2	96.0
B(1) + PVC(50t) + B(2)	56	222	149	67.1	66.2	95.5

* Detail Specifications of Materials are; refer to the Tables 3 and 4. In Table 3/4, A(1) CU 125NS prepreg, Vacuum bagging, 0.75 mm thick; A(2) CU 125NS prepreg, Vacuum bagging, 1.5 mm thick; A(3) CU 125NS prepreg, Vacuum bagging, 2.25 mm thick; B(1) CU 125NS(0.75 t) + CF 3327(1.5 t), Vacuum bagging, 2.25 mm thick; B(2) CU 125NS(2.25 t) + CF 3327(1.5 t), Vacuum bagging, 3.75 mm thick; C(1) CF 3327 fabric(6 plies), Infusion, 1.5 mm thick; C(2) CF 3327 fabric(12 plies), Infusion, 3.0 mm thick; C(3); CF 3327 fabric(18 plies), Infusion, 4.5 mm thick.

4. Design of CT Cradle

4.1. 3-Dimensional Design of Cradle

Figure 4 shows the 3-dimensional shape of the cradle. It consists of the region inserted headrest supported with the patient's head as shown in detail "A" of Figure 4a and the region fixed at the main frame, as shown in detail "B." The total length of the cradle is 2322 mm with a width of 465 mm. The core material of the sandwich cradle has the shape inserted in the headrest in the front, as shown Figure 4b, and the shape inserted aluminum reinforcement block for the part bolted through the hole.

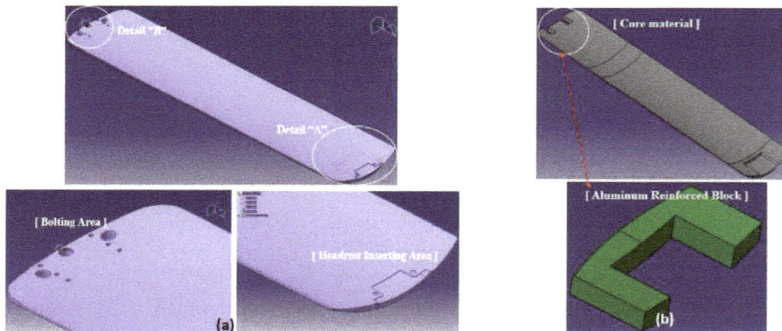

Figure 4. Shape of Sandwich composite Cradle; (**a**) 3-dimensional Shape of Cradle, (**b**) design of Core and the reinforced block.

4.2. Optimization of the Thicknesses of Face/Core Materials

Using the optimization process, the thicknesses of face and core materials for the sandwich cradle were determined to meet the deflection of end point for sandwich cradles were determined not more than 20 mm for the external 135 kg load and the X-ray transmission rate of sandwich cradle was equal to or greater than the X-ray transmission rate of 99.9% aluminum 1.5 mm thickness.

For optimization of the total thickness of the sandwich cradle, the total thickness (x1) of face material and the thickness (x2) of core material were considered as design variables. The objective function is used to determine the cost of material for the sandwich cradle and to determine x1 and x2 to minimize the cost. In general, the price of the face material is 250 US$/(1mm thick) per 1 m^2 and that of the core material is 1 US$/(1mm thick) per 1 m^2.

The cost of materials was calculated by:

$$\text{Area of cradle} = \text{Length} \times \text{width} = 2.65 \text{ m} \times 0.5 \text{ m} = 1.325 \text{ m}^2$$

$$\text{Cost} = f(x1, x2) = (\text{Area} \times 250 \text{ US\$}) \times (x1) + (\text{Area} \times 1 \text{ US\$}) \times (x2)$$

Therefore, the objective function is described in Equation (4);

$$\text{Objective function, } f(x1, x2) = 331.25 \times (x1) + 1.325 \times (x2) \tag{4}$$

Three conditions were considered to be the constraints for optimization: the limitation of deflection for the end point of sandwich cradle, the condition for X-ray transmission rate of sandwich cradle, and the design limitation for the total thickness of sandwich cradle. At first, the sandwich cradle required that the deflection of the end tip did not exceed 20 mm when a 135 kg load was applied as shown in Figure 5a. The total deflection (Δ) of the sandwich structure is defined in Equation (5) [21]

$$\Delta = \Delta 1 + \Delta 2 = \frac{WL^3}{48D} + \frac{WL}{4V} \tag{5}$$

where $D = E_f \frac{bt^3}{6} + E_f \frac{btd^2}{2} + E_c \frac{bc^3}{12} \cong E_f \frac{btd^2}{2}$, total thickness of face material x1 = 2 × t, thickness of face material = t, $V = AG$, $A = bd$, distance between each center of face material $d = \left(\frac{x1}{2}\right) + x2$, external load W = 135 kg = 1324.35 N, length of cradle L = 2650 mm, width of cradle b = 465 mm, Young's modulus of face material E_f = 50 GPa, Young's modulus of core material E_c = 36.7 MPa, and shear modulus of core material G = 6.4 MPa.

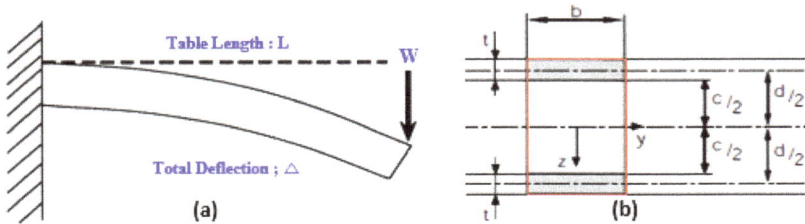

Figure 5. Structural Constraints for sandwich cradle; (**a**) deflection of sandwich cradle for external load, (**b**) cross-section and dimension notation of sandwich beam.

Equation (6) represents the first constraint, related to the total deflection and structural requirement.

$$G1(x1, x2) = \frac{WL^3}{48D} + \frac{WL}{4V} - 20 \leqq 0 \tag{6}$$

As the second constraint, the X-ray transmission rate must be greater than that of aluminum 1.5 mm thick ($\cong$ 75.22%).

$$G2(x1, x2) = 0.7522 - A(x1) \times B(x2) \leqq 0 \tag{7}$$

where A(x1) and B(x2) were derived from paragraphs 3.2 and 3.3. The relationship between X-ray transmission rate and total thickness of face material is represented by Equation (8).

$$A(x1) = \frac{-5.7752 \times x1 + 100.89}{100} \tag{8}$$

The relationship between X-ray transmission rate and the thickness of core material is represented by Equation (9).

$$B(x2) = \frac{-0.0822 \times x2 + 99.659}{100} \tag{9}$$

The third constraint is the limitation of the total thickness (T) of the sandwich cradle. Total thickness of the sandwich cradle must not exceed 50 mm. Therefore, this constraint is expressed by Equation (10).

$$G3(x1, x2) = x1 + x2 - 50 \leqq 0 \tag{10}$$

The optimization solution used in this study has been applied with linear programming (LP) using Microsoft Excel (Microsoft Corporation, Redmond, WA, USA). As shown in Figure 6, the optimization algorithm (1) suggests values for x1 and x2, (2) calculates deflection, X-ray transmission rate of sandwich cradle, and the total thickness of sandwich cradle, (3) repeats the process of verifying that the calculated values are met with constraint conditions, and (4) determines the x1 and x2 values with minimum material cost. The resulting total thickness (x1) of face and thickness (x2) of the core was decided to be 2.40 and 47.60 mm, respectively.

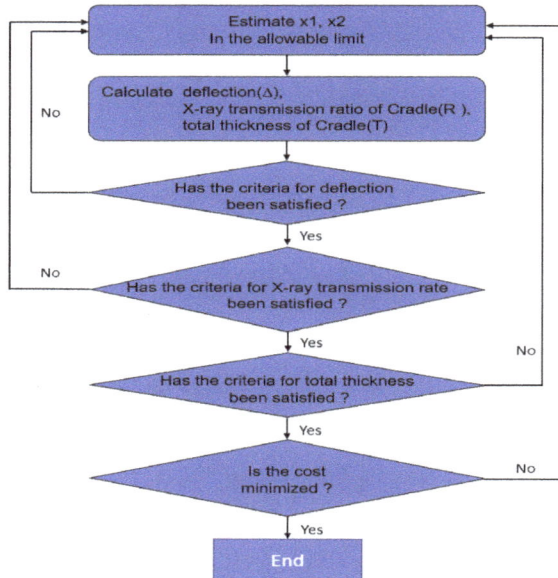

Figure 6. Optimization Algorithm for Sandwich Cradle.

4.3. Prediction of Stacking Sequences

The structural performance criteria of the cradle that the end point deflection should not exceed 20 mm when the total load applied at a specific location is 135 kg is shown in Figure 7a. Figure 7b,c indicate the boundary conditions of structural analysis. The finite element analysis (FEA) for prediction of the stacking sequence was performed using ANSYS Version 11 (Taesung S&E, Seoul, Korea). Using 3-D CAD data, the core was modeled by a solid element (Solid 95) and the facing materials were applied with layered shell elements (Shell 99). Each attribution, such as the proposed stacking patterns, material property, thickness, and stacking direction, were assigned to each portion of the facing material.

In Figure 8, the proposed stacking pattern is that CF 3327 (4 plies) is applied on the upper skin (Section A) and reinforced CF 3327 (2 plies) is applied in to the Section B area. CF 3327 (3 plies), MCU 125NS (3 plies) and CU 125 NS (3 plies) were proposed to be applied at Section D and were reinforced with CF 3327 (9 plies), MCU 125NS (3 plies), and CU 125 NS (19 plies), for a total of 31 plies at Section C for the lower skin. For all materials, the fiber orientation was laminated at 0 degrees. As the result of FEA, the end deflection was estimated as 19.02 mm, which satisfies the requirements as shown in Figure 9. As the result of FEA, the thickness of the X-ray transmission area was 2.65 mm and was similar to the thickness of the optimization results.

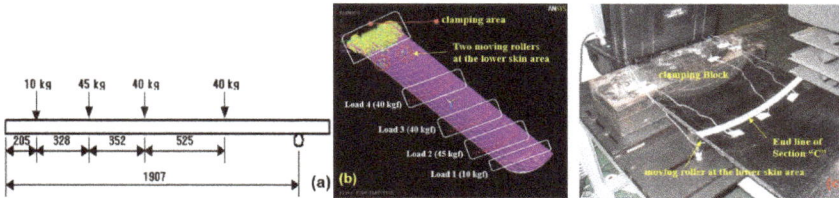

Figure 7. Loading conditions and Boundary conditions on structural analysis; (**a**) Loading conditions on the top surface, (**b**) Loading & Boundary conditions on ANSYS, (**c**) Real clamping condition.

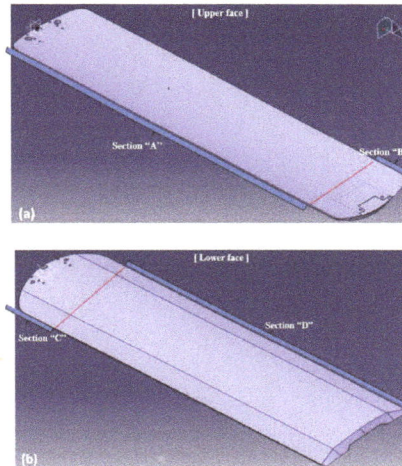

Figure 8. Proposed Sequence pattern; (**a**) Upper face (**b**) Lower face.

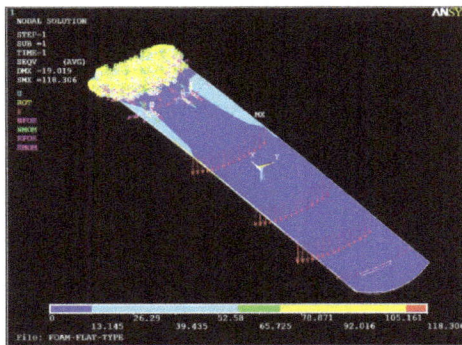

Figure 9. Result for structural analysis of sandwich cradle.

5. Manufacturing Process and Performance Evaluation of CT Cradle

5.1. Manufacturing Process of Cradle

The cradle designed as a sandwich structure is composed of the top face, bottom face, and the foam core between faces. The first layers of upper and lower face were formed by infusion and the remaining layers were formed using prepreg. The upper and lower faces were molded as shown in Figure 10a,b, after stacking carbon UD or carbon fabric as per the thickness estimated by (FEA). After applying additional resin on the cured face, the prepared core was inserted between two molds as shown in Figure 10c,d, and it was assembled by clamping. Then, the assembly was cured in a dry oven as shown in Figure 10e. After curing, the assembly was demolded, and the completed cradle was trimmed at the bonded area as shown in Figure 10f.

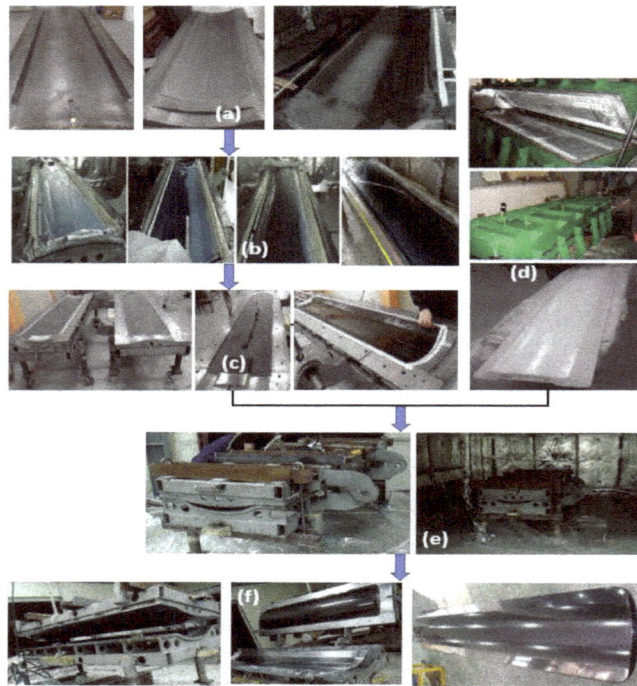

Figure 10. Manufacturing Process of Sandwich cradle; (**a**) preparing mold & layup of carbon fiber, (**b**) resin infusion, (**c**) curing the faces & apply the resin evenly, (**d**) prepared core, (**e**) inserting core into molds and curing/bonding, (**f**) de-molding and hole machining.

5.2. Stiffness Evaluation of Cradle

A stiffness test of the cradle determined whether it satisfied the criteria, which is that the end deflection is 20 mm or less when a total load of 135 kg was applied. The loading bar connected to the loadcell and LVDT (Linear Variable Differential Transformer) were equipped and the applied load and deflection at the prescribed point were measured as shown in Figure 11. As a result, the deflection at the end point of the cradle was found to be 19.49 mm. Therefore, it satisfied the criteria.

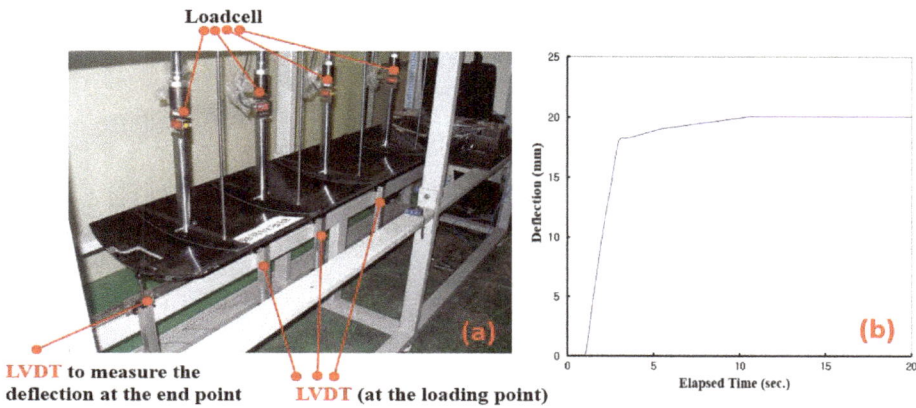

Figure 11. Real deflection test; (a) Test set-up, (b) Result of deflection test.

5.3. X-ray Transmission Performance of Cradle

The X-ray transmission rate for the final product was measured as shown in Figure 12 for the thickest part of the X-ray measurement area of the sandwich cradle. The conditions of X-ray transmission were 100 kV, 200 mA, and an exposure time of 0.1 s. The transmission rate of the sandwich cradle was 78.8%, which is better than the 75.22% transmission rate of the 1.5 mm thick aluminum (Table 6).

Uniform X-ray transmission performance of the sandwich cradle whose first ply for the upper face and lower faces was laid by infusion and the remaining layers used prepreg is shown in Figure 13b. This is in contrast to the usual defect as shown in Figure 13a.

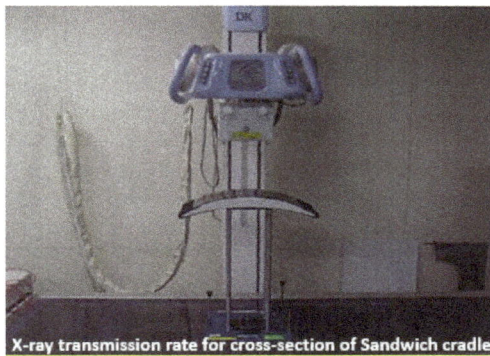

Figure 12. X-ray transmission test for thick section of sandwich cradle.

Table 6. X-ray Transmission rate of Final sandwich Cradle.

| | Measured Value (unit: mRem) | | | | X-ray Transmission Rate (%) |
	1	2	3	Average	$(I1/I0) \times 100$
I_0 (Initial value)	222.4	219.2	218.2	220.0	78.8
I_1 (sandwich cradle)	174.6	172.4	172.6	173.2	

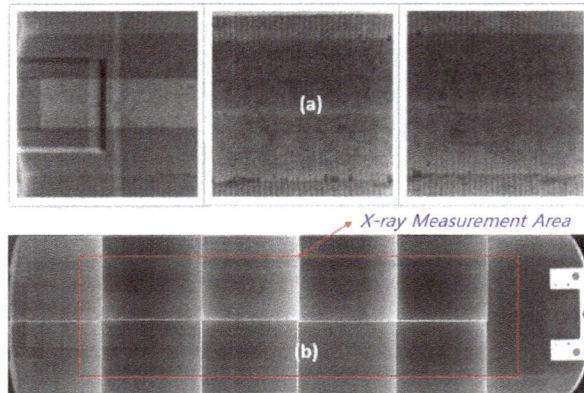

Figure 13. Uniform X-ray transmission of Composite Cradle; (**a**) bad case due to the void, excess resin, and poor resin of face and wrinkle of foam, (**b**) good case - 11 spit-shots of X-ray measurement area.

6. Conclusions

This paper evaluated the X-ray transmission rate for face, core and sandwich structure to determine the thickness of the sandwich cradle by repeatedly calculating the thickness of face and core materials to satisfy the structural performance, X-ray transmission conditions, and to propose the manufacturing process for the sandwich cradle.

1. The X-ray transmission performance of several face materials and some cases of core materials was measured. Measurements of the X-ray transmission rate of individual materials show that X-ray transmission performance was significantly affected by the density and thickness of the materials. The equations describing the relationship between thickness of face and core material and the X-ray transmission rate were derived through the measured value. X-ray transmission rates were measured for a sandwich structure combined with several cases of face and core, and the relation equation for X-ray transmission performance of sandwich structure was estimated.

2. The 3D shape of the sandwich cradle of CT instruments was decided with consideration to the bolting conditions and interface with other parts. The thicknesses of face and core materials were optimized to minimize the cost of the materials on meeting the requirements that the deflection of sandwich cradle should be not more than 20 mm and the X-ray transmission rate of cradle should be equal to or greater than that of aluminum 1.5 mm thick.

3. The first layers of upper and lower skin were formed by infusion and the remaining layers were used prepreg. After the upper and lower skins were molded, they were integrally bonded to the core and the sandwich cradle was formed. X-ray transmission performance of the final fabricated cradle was uniform over its entire surface. The X-ray transmission rate of the sandwich cradle is 78.8%, which is better than the 75.22% transmission rate of aluminum (1.5 mm thick). Additionally, it satisfied the criteria by finding the deflection at the end point of the cradle to be 19.49 mm.

4. The improvement of the CT cradle increases the accuracy of the medical device rather than the improvement of the CT system and an improvement in X-ray transmission rates can have an effect on reducing the amount of X-ray applied to patients.

Author Contributions: S.J.L. carried out the experimental study and wrote the manuscript. I.S.C. supervised the research and design of the study.

Funding: This research received no external funding.

Conflicts of Interest: The authors declare no conflict of interest.

Materials **2019**, *12*, 286

References

1. Kim, W.D.; Hong, D.J. Design of an aircraft composite window frame using VaRTM process. *Korean Soc. Compos. Mater.* **2006**, *19*, 1–7.
2. Shin, K.B.; Ko, H.Y.; Cho, S.H. A study on crashworthiness and rollover characteristics of low-floor bus made of honeycomb sandwich composites. *Korean Soc. Compos. Mater.* **2008**, *21*, 22–29.
3. Kim, J.S.; Jeong, J.C.; Lee, S.J. An experimental study on the hybrid composite carbody structure. *Korean Soc. Compos. Mater.* **2005**, *18*, 19–25.
4. Lee, S.J.; Kim, J.S.; Jeong, J.C.; Cho, S.H. A study for manufacturing process of train bodyshell with sandwich composite using the autoclave. *Jpn. J. Reinf. Plast.* **2006**, *52*, 269–277.
5. Lee, J.Y.; Shin, K.B.; Lee, S.J. A study on failure evaluation of Korean low floor bus structures made of hybrid sandwich composites. *Trans. KSAE* **2007**, *15*, 50–61.
6. Ning, H.; Janowski, G.M.; Vaidya, U.K.; Husman, G. Thermoplastic sandwich structure design and manufacturing for the body panel of mass transit vehicle. *Compos. Struct.* **2007**, *80*, 82–91. [CrossRef]
7. Ning, H.; Vaidya, U.; Janowski, G.M.; Husman, G. Design, manufacturing and analysis of a thermoplastic composite frame structure for mass transit. *Compos. Struct.* **2007**, *80*, 105–116. [CrossRef]
8. Garcea, S.C.; Wang, Y.; Withers, P.J. X-ray computed tomography of polymer composites. *Compos. Sci. Technol.* **2018**, *156*, 305–319. [CrossRef]
9. Mao, L.; Chiang, F. Mapping interior deformation of a composite sandwich beam using Digital Volumetric Speckle Photography with X-ray computed tomography. *Compos. Struct.* **2017**, *179*, 172–180. [CrossRef]
10. Rolfe, E.; Kelly, M.; Arora, H.; Hooper, P.A.; Dear, J.P. Failure analysis using X-ray computed tomography of composite sandwich panels subjected to full-scale blast loading. *Compos. Part B* **2017**, *129*, 26–40. [CrossRef]
11. Zhou, Y.; Zheng, Y.; Pan, J.; Sui, L.; Xing, F.; Sun, H.; Li, P. Experimental investigations on corrosion resistance of innovative stee-FRP composite bars using X-ray microcomputed tomography. *Compos. Part B* **2019**, *161*, 272–284. [CrossRef]
12. Bull, D.J.; Helfen, L.; Sinclair, I.; Sprearing, S.M.; Baumbach, T. A comparing of multi-scale 3D X-ray tomographic inspection techniques for assessing carbon fibre composite impact damage. *Compos. Sci. Technol.* **2013**, *75*, 55–61. [CrossRef]
13. Kolkoori, S.; Wrobel, N.; Zscherpel, U.; Ewert, U. A new X-ray backscatter imaging technique for non-destructive testing of aerospace mateirals. *NDT E Int.* **2015**, *70*, 41–52.
14. Revol, V.; Plank, B.; Kaufmann, R.; Kastner, J.; Kottler, C. Laminate fibre structure characterization of carbon fibre-reinforced polymers by X-ray scatter dark field imaging with a grating interferometer. *NDT E Int.* **2013**, *58*, 64–71.
15. Pinter, P.; Dietrich, S.; Bertram, B.; Kehrer, L.; Elsner, P.; Weidenmann, K.A. Comparison and error estimate of 3D fibre orientation analysis of computed tomography image data for fibre reinforced composites. *NDT E Int.* **2018**, *95*, 26–35.
16. Catapano, A.; Montemurro, M. A multi-scale approach for the optimum design of sandwich plates with honeycomb core. Part II: The optimisation strategy. *Compos. Struct.* **2014**, *118*, 677–690. [CrossRef]
17. Han, B.; Qin, K.; Yu, B.; Zhang, Q.; Chen, C.; Lu, T.J. Design optimization of foam-reinforced corrugated sandwich beams. *Compos. Struct.* **2015**, *130*, 51–62. [CrossRef]
18. An, H.; Chen, S.; Huang, H. Optimal design of composite sandwich structures by considering multiple structure cases. *Compos. Struct.* **2016**, *152*, 676–686. [CrossRef]
19. Ikeya, K.; Shimoda, M.; Shi, J. Multi-objective free-form optimization for shape and thickness of shell structures with composite materials. *Compos. Struct.* **2016**, *135*, 262–275. [CrossRef]
20. Nasab, F.F.; Geijselaers, H.J.M.; Baran, I.; Akkerman, R.; de Boer, A. A level-set-based strategy for thickness optimization of blended composite structures. *Compos. Struct.* **2018**, *206*, 903–920. [CrossRef]
21. Allen, H.G. *Analysis and Design of Structural Sandwich Panels*, 1st ed.; Pergamon Press: Oxford, UK, 1969; pp. 8–18.

materials

MDPI

Article

Fatigue Behavior of Concrete Beam with Prestressed Near-Surface Mounted CFRP Reinforcement According to the Strength and Developed Length

Hee Beom Park, Jong-Sup Park, Jae-Yoon Kang and Woo-Tai Jung *

Structural Engineering Research Institute, Korea Institute of Civil Engineering and Building Technology, Goyang 10223, Korea; heebeompark@kict.re.kr (H.B.P.); jspark1@kict.re.kr (J.-S.P.); jykang@kict.re.kr (J.-Y.K.)
* Correspondence: woody@kict.re.kr; Tel.: +82-31-910-0580

Received: 25 November 2018; Accepted: 21 December 2018; Published: 24 December 2018

Abstract: The prestressed near-surface mounted reinforcement (NSMR) using Fiber Reinforced Polymer (FRP) was developed to improve the load bearing capacity of ageing or degraded concrete structures. The NSMR using FRP was the subject of numerous studies of which a mere portion was dedicated to the long-term behavior under fatigue loading. Accordingly, the present study intends to examine the fatigue performance of the NSMR applying the anchoring system developed by Korea Institute of Construction and Building Technology (KICT). To that goal, fatigue test is performed on 6.4 m reinforced concrete beams fabricated with various concrete strengths and developed lengths of the Carbon Fiber Reinforced Polymer (CFRP) tendon. The test results reveal that the difference in the concrete strength and in the developed length of the CFRP tendon has insignificant effect on the strengthening performance. It is concluded that the accumulation of fatigue loading, the concrete strength and the developed length of the tendon will not affect significantly the strengthening performance given that sufficient strengthening is secured.

Keywords: CFRP; fatigue; prestressed near-surface mounted reinforcement (NSMR); strengthening; tendon

---

1. Introduction

Prestressed concrete (PSC) eases the control of the deflection and cracks in concrete structures by reducing the tensile stress through the introduction of a compressive force by means of tendons embedded in the tension zone of the structure. Ageing bridges may experience loss of their function and performance that should be recovered or improved by strengthening the structure. The strengthening of PSC girder bridge is achieved by enlarging the girder, external prestressing, carbon fiber bonding or steel plate bonding. The external bonding reinforcement (EBR) methods were applied to improve the performance by bonding the reinforcement on the tensile zone of the concrete member using an adhesive. This reinforcement took first the form of a plate made of steel that started to be replaced by Fiber Reinforced Polymer (FRP) since the mid of 1980s.

The near-surface mounted reinforcement (NSMR) resembles the above mentioned external bonding reinforcement but embeds the FRP plate or bar in a groove cut with a definite depth in the concrete surface. De Lorenzis et al. [1] reported that the superior resistance to bond failure provided by NSMR compared to EBR brought higher improvement of the performance and that the embedment of the tendon in concrete could prevent its damage due fire or vehicle impact. Moreover, these authors also reported that additional resistance to physical wearing caused by the tires of passing vehicles can be expected when NSMR is applied to strengthen the negative moment zone of the deck.

In view of the efficiency of FRP, both NSMR and EBR cannot exploit fully the maximum performance of FRP due to the premature occurrence of debonding at the reinforcement-concrete interface. As passive strengthening methods, their effect appears only under the application of additional live loads without clear improvement of the serviceability in term of cracking and deflection recovery. Accordingly, numerous researchers attempted the prestressed NSMR achieving the synergy of both NSMR and prestressing [2]. For the prestressed NSMR to be realizable, the most important matter is the development of appropriate anchor and tensioning system enabling to apply the prestress force using FRP. El-Hacha and Soudki [2] employed temporary clamp-anchors and pre-tensioning or end-supported tensioning devices. Korea Institute of Construction and Building Technology (KICT) developed a tensioning device using anchors fixed to the concrete beam as tensile reaction table [3].

In particular, FRP is checked to apply in many sites because of strong points such as high strength, light weight, chemical resistance and need to secure reliability through real size test because of low applications, low construction cases, problems with design criteria [4–8].

Despite of the numerous studies related to NSMR using FRP, a very few of them studied the long-term behavior under fatigue loading [9]. The repeated action of the fatigue loading is likely to degrade the quality of FRP and concrete and the concrete-FRP-filler bond performance in the structure strengthened by NSMR. A few studies have been published on RC beams strengthened by the NSM FRP system with epoxy adhesive and subjected to flexural fatigue loading [10–15].

Accordingly, the present study intends to examine the fatigue behavior of the prestressed NSMR applying the anchoring device developed by KICT. The cumulated effect of the fatigue load on the strengthening performance is examined by means of a series of fatigue tests performed on 6.4 m reinforced concrete beams fabricated with various concrete strengths and developed lengths of the Carbon Fiber Reinforced Polymer (CFRP) tendon. The fatigue tests were conducted in two stages. The first stage applied 2 million loading cycles and the second stage applied static loading until failure of the specimens to measure the residual strength after the accumulation of fatigue.

2. Experimental Program

2.1. Test Variables

This study intends to examine the fatigue performance of the prestressed NSMR. To that goal, the specimens are designated as shown in Figure 1 with respect to the concrete strength and developed length of the CFRP tendon chosen as test variables. In Figure 1, the first string of character indicates the type of filler with E for epoxy. The second string of character announces the type of surface treatment applied on the FRP tendon with OX for oxide-coating. The third string of character designates the amount of reinforcement with 1 for one line of reinforcement. The fourth string of character stands for the concrete strength with 30 for 30 MPa and 40 for 40 MPa. The fifth string of character indicates the developed length of the CFRP tendon. Specifically, for the span length of 6000 mm, 67% represents 4000 mm and 93% represents 5600 mm with reference to the length of 4800 mm corresponding to about 80% of the span length. Table 1 lists the adopted specimens with their designation according to the test variables. Steel rebars had a modulus of elasticity of 200 GPa and a yielding stress of 400 MPa. Epoxy had a tensile strength of 47 MPa and a bond strength of 9 MPa.

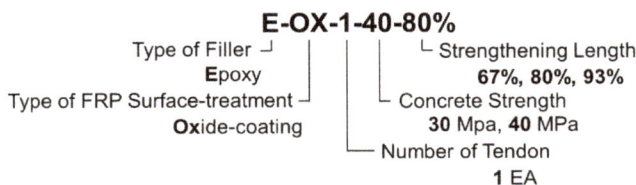

E-OX-1-40-80%

Type of Filler ⌐   ⌐ Strengthening Length
Epoxy   **67%, 80%, 93%**
Type of FRP Surface-treatment ⌐   ⌐ Concrete Strength
Oxide-coating   **30 Mpa, 40** MPa
    ⌐ Number of Tendon
    1 EA

Figure 1. Designation of specimens.

Table 1. Test variables and designation of specimens.

Name	Type of Filler	Type of FRP Surface-Treatment	Number of Tendon (EA)	f_{ck} (MPa)	Strengthing Length (mm)
Control	-	-	-	40	-
E-OX-1-30-80%	Epoxy	Oxide-coating	1	30	4800
E-OX-1-40-67%	Epoxy	Oxide-coating	1	40	4000
E-OX-1-40-80%	Epoxy	Oxide-coating	1	40	4800
E-OX-1-40-93%	Epoxy	Oxide-coating	1	40	5600

2.2. Fabrication of Specimens

As shown in Figure 2a, the reinforced concrete specimens were fabricated to have a total length of 6400 mm with a span length of 6000 mm. Figure 2b depicts the reinforcement details of the specimens. Three D19 bars are used as tensile reinforcement and D22 bar is disposed in the compression zone to prevent the occurrence of compressive failure before tensile failure. Moreover, D10 stirrups are installed every 200 mm to provide sufficient resistance to shear. For the prestressed NSMR, grooves must be cut for the anchoring of the FRP tendon to be embedded in concrete. As shown in Figure 2c, the grooves are rectangular with width of 30 mm and depth of 40 mm. The surface is finished by applying epoxy after the tensioning.

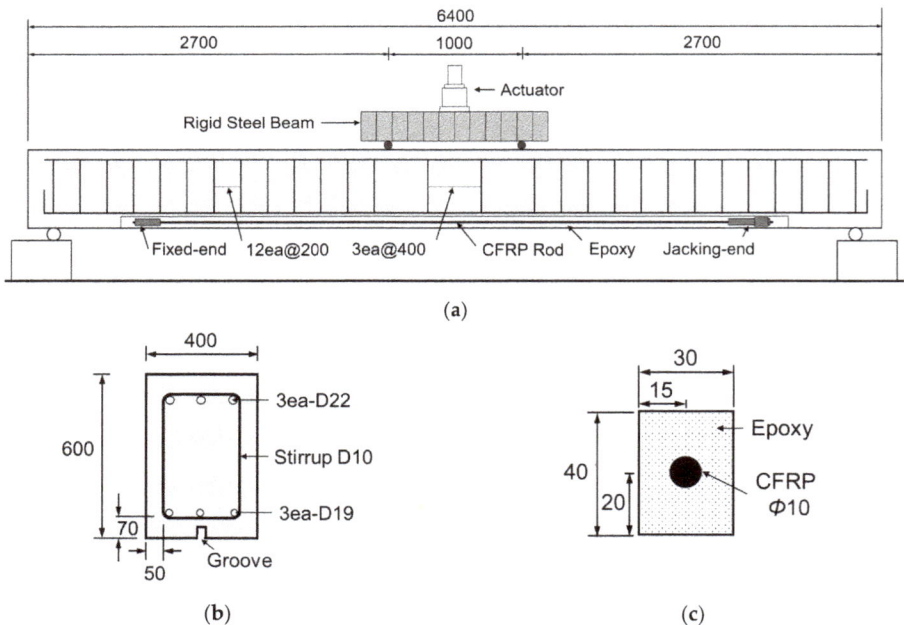

(a)

(b)

(c)

Figure 2. Test setup and beam details (all dimensions are in mm): (**a**) beam dimensions; (**b**) cross section; (**c**) groove.

2.3. Test Setup

The structural tests were executed in two stages. The first stage applied 2 million loading cycles and the second stage applied static loading until failure of the specimens to measure the residual strength after the accumulation of fatigue.

The size of the fatigue load was determined with reference to the stress of the tensile member using the calculation method suggested by ACI, AASHTO and CSA [16–19]. CSA limits the range of the tensile member stress below 125 MPa, and ACI limits the size of the fatigue load within 80% of the yield strength. Based on these recommendations, Oudah and El-Hacha [20] conducted their experiments with a fatigue load ranging between $0.42f_y$ and $0.7f_y$, where f_y is the steel yield stress. Accordingly, the present study adopted fatigue load ranging between 60 kN and 100 kN. In addition, the loading rate was set to 2.0 to 3.0 Hz. The fatigue test conducted using a dynamic actuator with capacity of 1000 kN to apply 2 million loading cycles. Static load of 100 kN was applied after 1, 1000, 5000, 10,000, 100,000, 1 million and 2 million cycles to measure the deflection, strain and cracks and examine the eventual progress of damage according to the accumulation of fatigue.

Static loading was performed by 4-point loading on all the specimens using static actuators (Korea Institute of Civil Engineering and Building Technology, Goyang-Si, Korea) with capacity of 2000 kN to measure the deflection and strain according to the gradual increase of the load until failure. As shown in Figure 2a, loading was applied identically on all the specimens at the points located 500 mm far from the center to secure a pure flexure section of 1 m. The static loading was conducted through displacement control at speed of 0.03 mm/s for the first 20 mm and at speed of 0.05 mm/s beyond that displacement and until failure. Rebar strain gauges were attached to the upper and lower reinforcements to measure the rebar strain according to the load. Strain gauges were also attached at the center and quarter points of the CFRP tendon to grasp the strain change throughout the tests. As shown in Figure 3, the load-displacement curves were measured by reading directly the load of the actuator and the displacement from a LVDT installed at mid-span of the specimens.

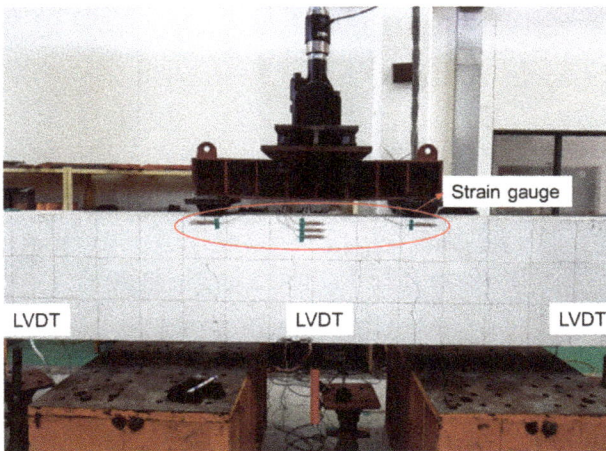

Figure 3. Sensor setup.

2.4. Tensioning

The tendon used in the fabrication of NSM specimens is a CFRP rod with diameter of 10 mm and its mechanical characteristics are arranged in Table 2. The allowable tension force to be introduced in the proposed FRP prestressed NSMR ranges between 40% and 65% of the strength of the FRP tendon [16]. Accordingly, the tension force of 100 kN corresponding to approximately 42% of the tendon strength was applied in the fabrication of the specimens. The tension force was measured by means of the load cell attached to the hydraulic cylinder and verified using the sensors disposed on the FRP tendon. Figure 4 plots the tension force measured the tensioning of the prestressed NSMR. Note that the strain of the CFRP tendon ranged between 6000×10^{-6} and 6500×10^{-6} during the tensioning work.

Table 2. Physical properties of 10-mm FRP tendon.

Failure Load (kN)	Tensile Strength (MPa)	Median Strain at Rupture (10^{-6})	Elastic Modulus (GPa)
236	3000	17,000	185

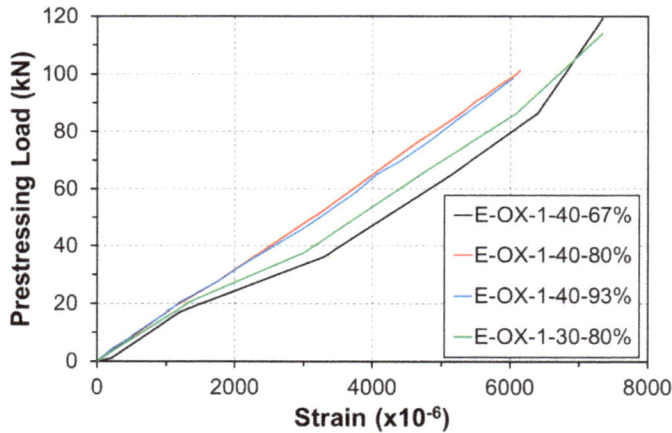

Figure 4. CFRP prestressing load wrt stress-strain curve.

3. Test Rsults

3.1. Fatigue Performance Acording to Concrete Srength

Table 3 arranges the values of the deflection at mid-span, the strains at the center of the upper and lower rebars, and the strain change of the CFRP tendon according to strength of the concrete and measured in each specimen under static load of 100 kN after 1, 1000, 2 million fatigue loading cycles. The increase of the deflection at mid-span following the accumulation of fatigue is seen to be provoked by the concrete creep, the degradation of the material bonding the tendon and concrete, and the plastic deformation of the CFRP tendon [21,22]. As shown in Figure 5, the deflection experienced the largest increase rate below 1000 loading cycles and increased gradually until 100,000 cycles to remain practically constant until 2 million cycles.

Table 3. Fatigue test results at upper load limit (Concrete strength).

Beam	Cycle	Deflection (mm)	Change (%)	Strain at Mid-Span					
				Top Steel		Bottom Steel		CFRP Rod	
				Strain (10^{-6})	Change (%)	Strain (10^{-6})	Change (%)	Strain (10^{-6})	Change (%)
Control	1	11.34	-	−181	-	1378	-	N/A [a]	-
	1000	12.98	14	−187	3	1561	13	N/A	N/A
	2,000,000	14.63	29	−222	23	1820	32	N/A	N/A
E-OX-1-30-80%	1	6.26	-	−212	-	687	-	1135	-
	1000	7.08	13	−194	−8	857	25	1290	14
	2,000,000	11.92	90	−283	33	1088	58	1360	20
E-OX-1-40-80%	1	5.96	-	−242	-	687	-	1001	-
	1000	6.88	15	−251	4	827	20	1167	17
	2,000,000	9.22	55	−274	13	1071	56	1258	26

[a] Control test specimen is not applied CFRP Rod.

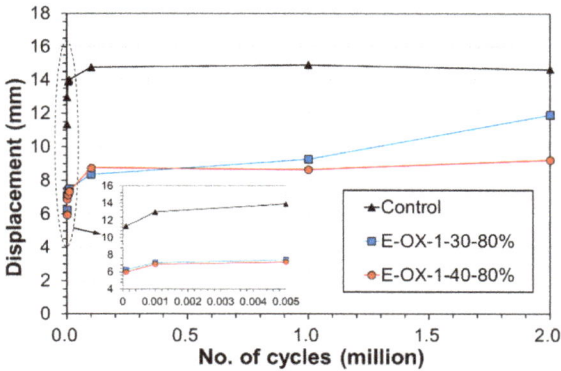

Figure 5. Variation of mid-span deflection during fatigue loading at upper load limit.

The residual strength can be measured using the static loading results listed in Table 4. Figure 6a plots the load-deflection curves measured at mid-span. The comparison of the residual strength with respect to the concrete strength reveals that, compared to Control, specimen E-OX-1-30-80% developed yield strength larger by 42% and ultimate strength larger by 51%. Similar observation can also be done for specimen E-OX-1-40-80%.

Table 4. Experimental results of NSM-strengthened RC beam (Concrete strength).

Beam	Yield Load (kN)	Ultimate Load (kN)	Change (%)	
			Yield Load	Ultimate Load
Control	142.91	183.84	-	-
E-OX-1-30-80%	202.97	277.69	42	51
E-OX-1-40-80%	198.51	276.68	39	51

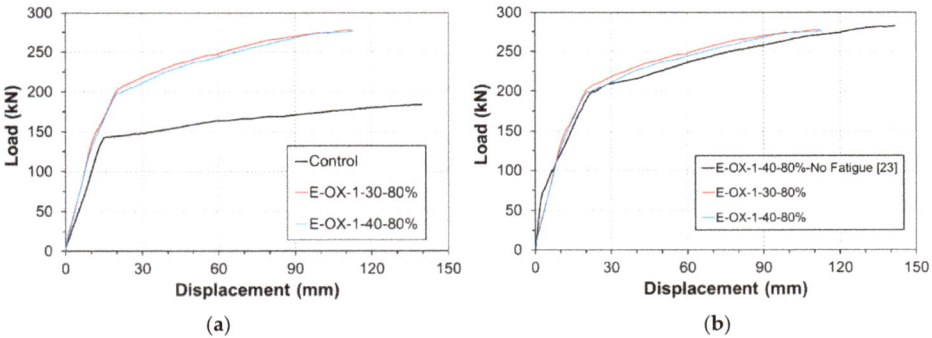

(a)

(b)

Figure 6. Load-displacement relationships (Concrete strength): (**a**) vs. control; (**b**) vs. no fatigue.

Figure 6b shows the residual strength of the specimens after fatigue loading. The results are seen to be nearly identical to those obtained in a previous work [23] in which static loading was applied without preliminary fatigue loading. When static loading is applied without accumulation of fatigue, the load-deflection curve presents three distinct behaviors that are the crack behavior, the yield behavior and the fatigue behavior. However, the load-deflection curve of the specimen which experienced fatigue loading presents two behaviors that are the yield behavior and the fatigue behavior without crack behavior since the specimen has already cracked. This indicates that damage or loss of the performance did practically not occur following the accumulation of fatigue. Moreover, there was no significant difference in the performance developed by the specimens with concrete strength of 30 MPa and 40 MPa. This means that the strengthening effect is not particularly influenced by the accumulation of fatigue nor the concrete strength if sufficient strengthening is secured.

Figure 7 plots the strains measured at the center and quarter lengths of the CFRP tendon under the upper load according to the accumulation of fatigue loading cycle. As shown in Figure 7a, similarly to the deflection, the strain exhibited its largest increase rate below 1000 loading cycles and increased gradually until 100,000 cycles to remain nearly unchanged until 2 million cycles. The strain at early fatigue would have decreased in occurrence of slip between the epoxy and the tendon [24] but such behavior cannot be observed here and resembles that of the deflection, which indicates that bond slip did not occur. Recalling that the CFRP tendon is 4800 mm-long, the strains plotted in Figure 7b were measured at the center (2400 mm) and quarter lengths (1200 mm and 3600 mm). In the graph, the horizontal axis represents the position along the length of the CFRP tendon and the vertical axis represents the strain at upper load after 2 million cycles of fatigue loading. The increase of the strain at the center of the tendon can be clearly distinguished with the increase of the load because the CFRP tendon-concrete bond performance is secured by the epoxy filler. This indicates the absence of bond slip in the anchored ends according to the accumulation of fatigue.

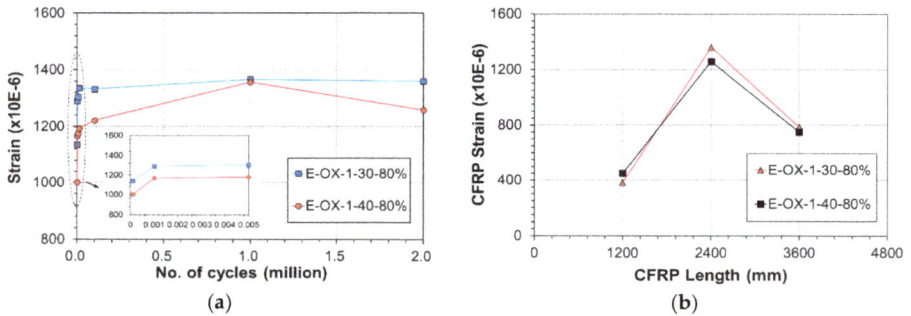

Figure 7. Variation of CFRP strain during fatigue loading at upper load limit: (**a**) mid span; (**b**) location.

Figure 8 plots the strain of the tendon under static loading. The indicated strain gathers the strain developed during tensioning and the strain developed along the accumulation of fatigue. Both specimens are seen to show similar strains and behavior at failure of the tendon. In addition, the tendon reached its tensile performance before rupture under the loading applied after the compressive failure of concrete. Consequently, the accumulation of the fatigue load appears to have a poor effect on the loss of the performance or tension force of the CFRP tendon.

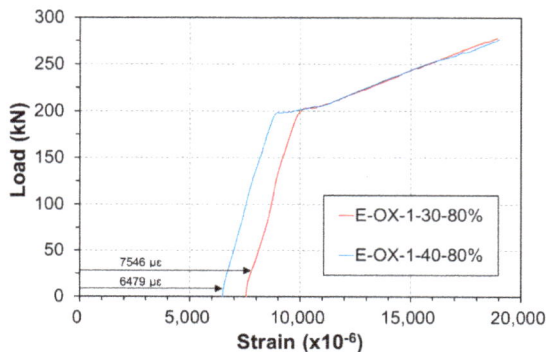

Figure 8. Load-strain relationship at CFRP (Concrete strength).

Figure 9 describes the variation of the strain at mid-span of the beam at upper load according to the accumulation of fatigue loading. The strain in all the specimens experienced steep increase from 1 cycle to 1000 cycles and stabilized gradually until 2 million cycles. In some case, the strain remained unchanged or decreased beyond 1 million cycles.

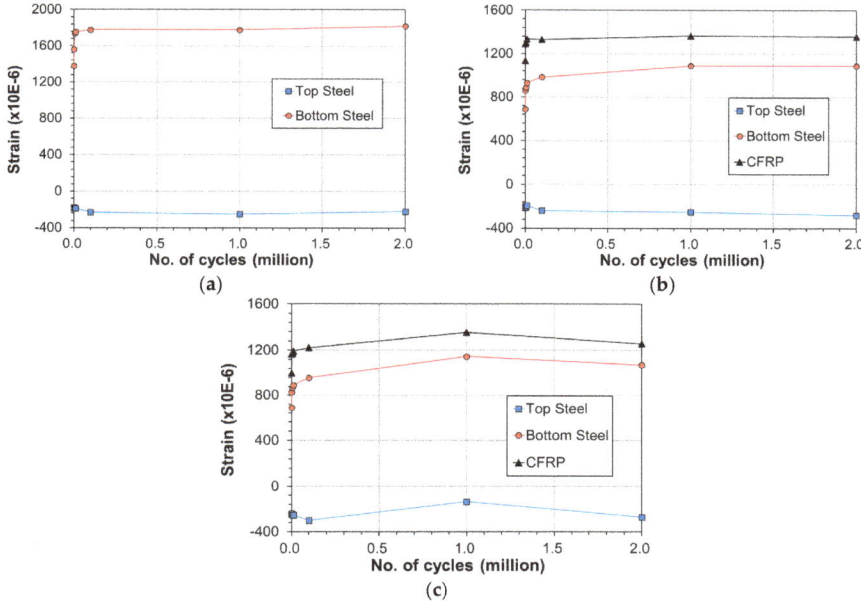

Figure 9. Variation of strain across the beam at mid-span with number of cycles: (**a**) control; (**b**) E-OX-1-30-80%; (**c**) E-OX-1-40-80%.

The ductility of the specimens with respect to the concrete strength is shown in Figure 10. Here, the ductility is defined as the ratio of the deflection at yielding to that at failure. The ductility of the prestressed NSMR specimens tends to decrease compared to that before strengthening. A previous study reported that the ductility tended to reduce by about 45% when prestressed NSMR is achieved using a CFRP tendon with high bond strength [18]. This is confirmed here with a decrease of the ductility by 42% for specimen E-OX-1-30-80% and by 40% for specimen E-OX-1-40-80% compared to Control and, indicates that the accumulation of fatigue loading does not affect significantly the ductility.

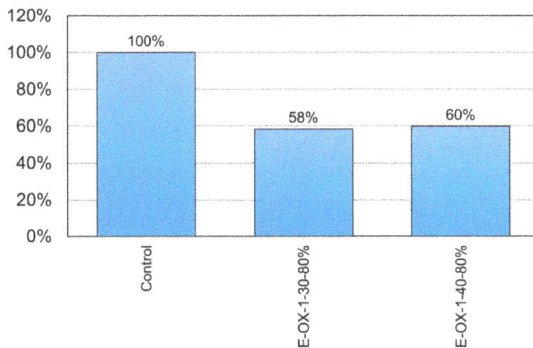

Figure 10. Ductility of NSM specimens (Concrete strength).

3.2. Fatigue Performance According to Developed Length of CFRP Tendon

Table 5 arranges the deflection at mid-span, the strain in the upper and lower reinforcements at mid-span and the strain change at the center of the CFRP tendon measured under static load of 100 kN after 1, 1000 and 2 million fatigue cycles and according to the developed length of the CFRP tendon. As shown in Figure 11, the deflection exhibited the largest increase rate below 1000 cycles, increased gradually until 100,000 cycles and stabilized until 2 million cycles. Compared to Control, the deflection under upper load after 2 million cycles was smaller by 44% for specimen E-OX-1-40-67%, by 37% for specimen E-OX-1-40-80% and by 22% for specimen E-OX-1-40-93%.

Table 5. Fatigue test results at upper load limit (Bond length).

Beam	Cycle	Deflection (mm)	Change (%)	Strain at Mid-Span					
				Top Steel		Bottom Steel		CFRP Rod	
				Strain (10^{-6})	Change (%)	Strain (10^{-6})	Change (%)	Strain (10^{-6})	Change (%)
Control	1	11.34	-	−181	-	1378	-	N/A [a]	-
	1000	12.98	14	−187	3	1561	13	N/A	N/A
	2,000,000	14.63	29	−222	23	1820	32	N/A	N/A
E-OX-1-40-67%	1	5.76	-	−195	-	493	-	907	-
	1000	7	22	−197	1	660	34	1084	20
	2,000,000	8.14	41	−253	30	864	75	1225	35
E-OX-1-40-80%	1	5.96	-	−242	-	687	-	1001	-
	1000	6.88	15	−251	4	827	20	1167	17
	2,000,000	9.22	55	−274	13	1071	56	1258	26
E-OX-1-40-93%	1	7.64	-	−230	-	813	-	1053	-
	1000	9.02	18	−221	-4	979	20	1212	15
	2,000,000	11.36	49	−203	-12	1280	57	1397	33

[a] Control test specimen is not applied CFRP Rod.

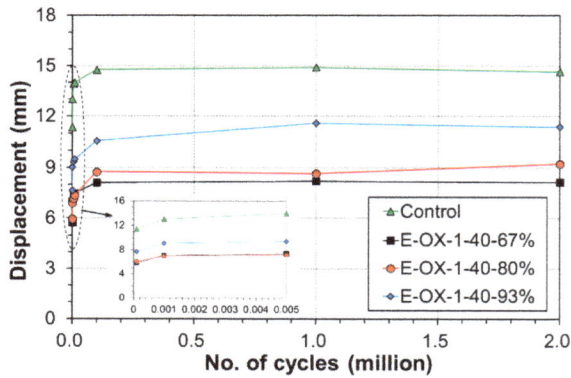

Figure 11. Variation of mid-span deflection during fatigue loading at upper load limit.

Table 6 arranges the results of the static loading test for the residual strength with respect to the developed length of the CFRP tendon. The load-deflection curves at mid-span of the beam members plotted in Figure 12a show that, compared to Control, specimen E-OX-1-40-67% developed yield load larger by 45% and ultimate load larger by 54% and that specimen E-OX-1-40-80% also developed comparable residual strength. As shown in Figure 12b, the residual strength after fatigue loading is practically identical to that developed by specimens which were subjected only to static loading. This indicates that the accumulation of fatigue does not provoke any damage nor performance loss of the prestressed NSMR specimens. In addition, the insignificance of the change in the performance developed by the specimens with developed lengths of 67%, 80% and 93% of the CFRP tendon demonstrates that the developed length of the tendon has practically no effect on the strengthening performance given that appropriate tensioning has been secured. However, shear failure would occur due to the increase of inclined cracks if the anchors are installed outside the effective depth d [3]. This is the case for specimen E-OX-1-40-67% in which the anchors are disposed outside the effective depth d and inclined cracks are seen to occur around the anchor as shown in Figure 13. Following, an appropriate and sufficient developed length should be secured to prevent the anchor be installed outside the effective depth d.

Table 6. Experimental results of NSM-strengthened RC beam (Bond Length).

Beam	Yield Load (kN)	Ultimate Load (kN)	Change (%)	
			Yield Load (kN)	Ultimate Load (kN)
Control	142.91	183.84	-	-
E-OX-1-40-67%	207.51	283.31	45	54
E-OX-1-40-80%	198.51	276.68	39	51
E-OX-1-40-93%	207.71	283.25	45	54

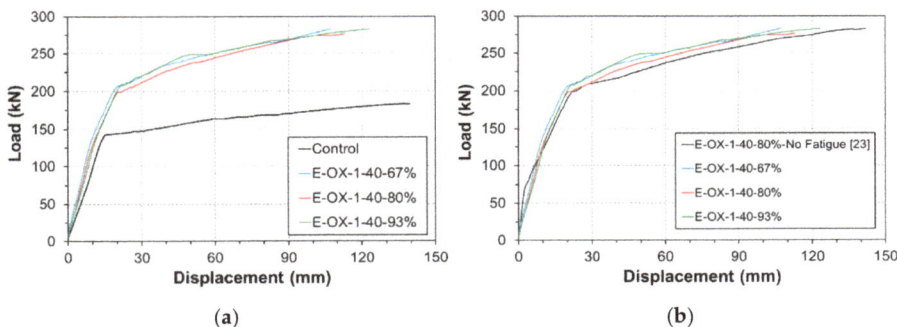

Figure 12. Load-displacement relationships (Bond length): (**a**) vs. control; (**b**) vs. no fatigue.

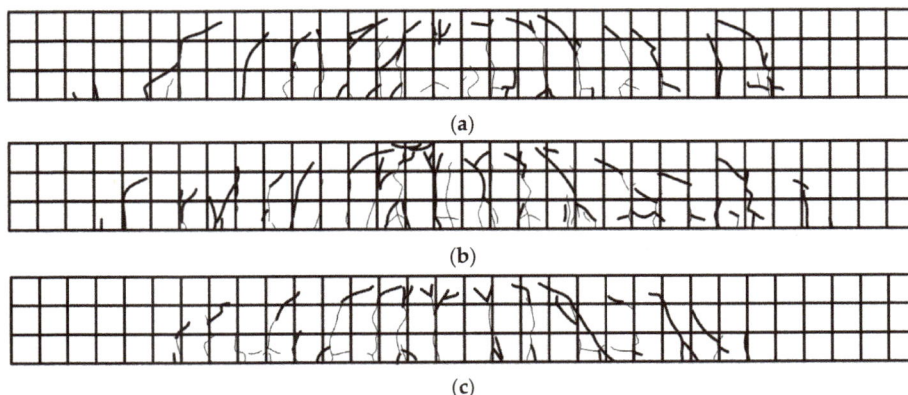

Figure 13. Crack map: (**a**) E-OX-1-40-67%; (**b**) E-OX-1-40-80%; (**c**) E-OX-1-40-93%.

Figure 14a plots the strains measured at the center and quarter lengths of the CFRP tendon under the upper load according to the accumulation of fatigue loading. Similarly to the deflection behavior, the strain exhibited the largest increase rate below 1000 cycles, increased gradually until 100,000 cycles and stabilized until 2 million cycles. As mentioned above, this indicates that bond slip did not occur at the epoxy-tendon interface. The strains were measured at lengths of 4000 mm, 4800 mm and 5600 mm along the tendon with reference to the center at the same positions listed in Table 7. Figure 14b plots the strains at upper load measured after 2 million cycles. Here also, it appears that bond slip did not occur at the anchored ends with respect to the length of the tendon.

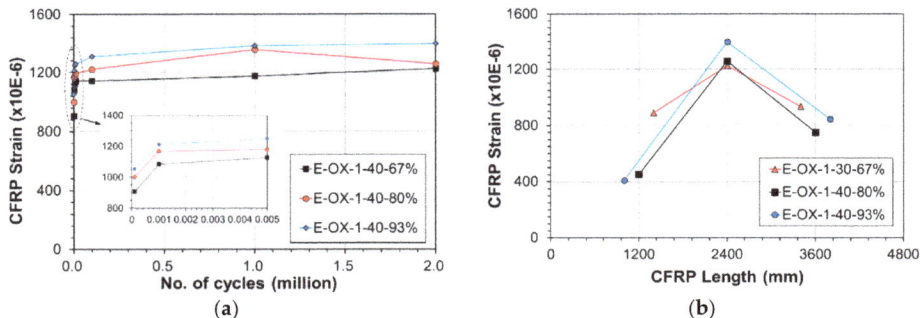

Figure 14. Variation of CFRP strain during fatigue loading at upper load limit: (**a**) mid span; (**b**) location.

Table 7. Strain gauge location.

Specimen	Strain Gauge Location (mm)		
	Quarter	Center	Quarter
E-OX-1-40-67%	1400	2400	3400
E-OX-1-40-80%	1200	2400	3600
E-OX-1-40-93%	1000	2400	3800

Figure 15 presents the strain of the tendon according to the static load. The indicated strains include the strain developed at tensioning and the strain caused by the accumulated fatigue loading. All the three specimens exhibit similar behavior and strain at failure. Moreover, the tendon reached its tensile performance before rupture under the loading applied after the compressive failure of concrete. Accordingly, the accumulation of the fatigue load appears to have poor effect on the loss of the performance like the tension force of the CFRP tendon.

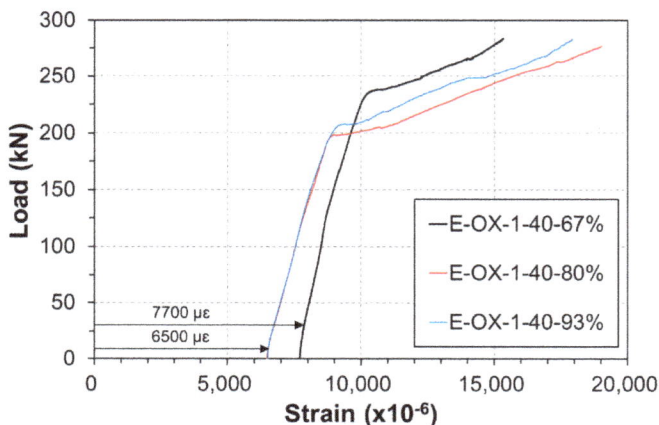

Figure 15. Load-strain relationship at CFRP (Bond length).

Figure 16 describes the steel and CFRP strain variation at mid-span of the beam members under the upper load according to the accumulation of fatigue loading. The strain in all the specimens is seen to experience the largest increase rate below 1000 cycles, to increase gradually until 100,000 cycles and to stabilize until 2 million cycles. In some case, the strain remained unchanged or decreased beyond 1 million cycles.

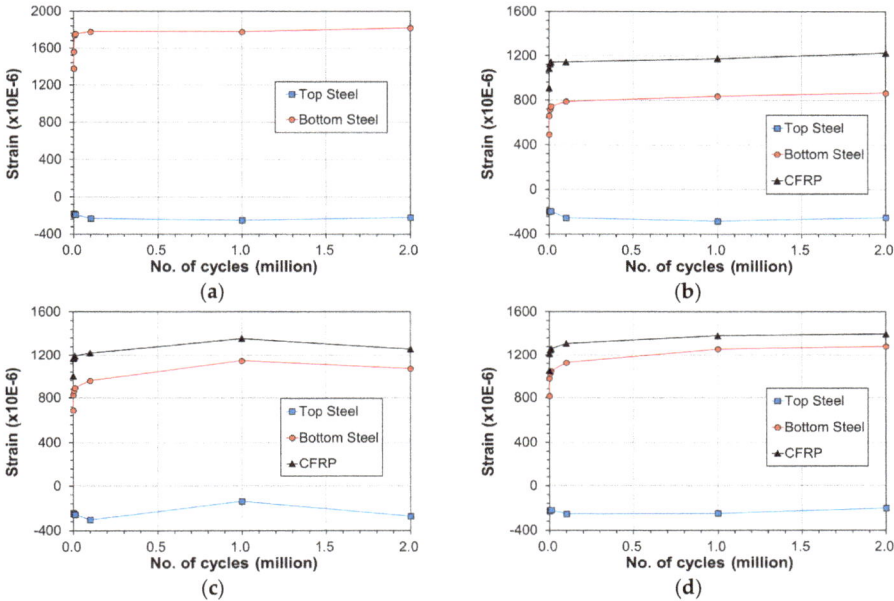

Figure 16. Variation of strain across the beam at mid-span with number of cycles: (**a**) control; (**b**) E-OX-1-40-67%; (**c**) E-OX-1-40-80%; (**d**) E-OX-1-40-93%.

Figure 17 shows the ductility of the specimens with respect to the developed length of the CFRP tendon. The ductility appears to have reduced by 44% for specimen E-OX-1-40-67%, by 40% for specimen E-OX-1-40-80% and by 38% for specimen E-OX-1-93% compared to Control. In other words, the ductility increased with longer developed length of the tendon. Therefore, longer developed length of the tendon seems to be favorable for securing stable ductile behavior. Moreover, since the observed ductility complies with that reported in a previous study [25], it can be stated that the accumulation of fatigue loading has no particular effect on the ductility of the prestressed NSMR specimens with respect to the developed length.

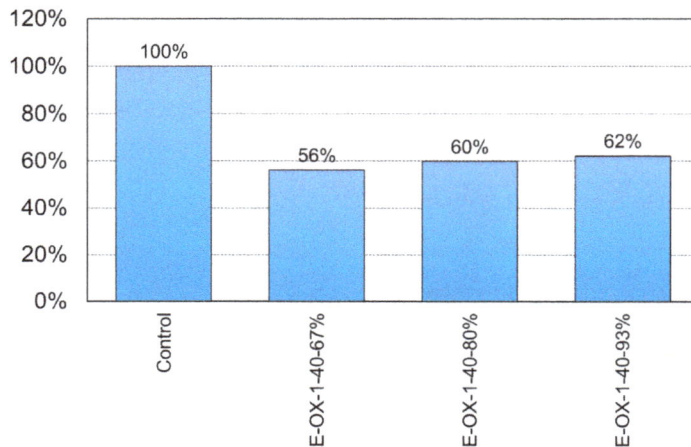

Figure 17. Ductility of NSM specimens (Bond length).

4. Conclusions

This study examined the strengthening performance of the prestressed NSMR according to the accumulation of fatigue loading. To that goal, fatigue test was performed on 6.4 m reinforced concrete beams fabricated with various concrete strengths and developed lengths of the CFRP tendon. The fatigue tests were conducted in two stages. The first stage applied 2 million loading cycles and the second stage applied static loading until failure of the specimens to measure the residual strength after the accumulation of fatigue. The following conclusions can be derived.

- The observation of the fatigue behavior with respect to the concrete strength revealed that the deflection exhibited the largest increase rate below 1000 cycles, increased gradually until 100,000 cycles and stabilized until 2 million cycles. The strain measured at the center of the CFRP tendon showed also similar tendency according to the accumulation of fatigue cycles. Moreover, the same tendency was also observed for the fatigue behavior according to the developed length of the tendon.

- The analysis of the strain developed during tensioning and the strain caused by the accumulation of fatigue revealed that all the specimens presented similar behavior and strain at the rupture of the tendon. In addition, the tendon could reach its tensile performance before rupture under the loading applied after the compressive failure of concrete. This indicated that the accumulation of the fatigue load had poor effect on the loss of the performance like the tension force of the CFRP tendon.

- The results of the static loading test with respect to the concrete strength showed that the accumulation of the fatigue load did not provoke damage nor performance loss of the prestressed NSMR specimens. Moreover, the test results of the specimens with concrete strengths of 30 MPa and 40 MPa appeared to be practically identical to those of a previous experiment without fatigue load accumulation. This indicated that the strengthening performance of the prestressed NSMR is insensitive to the accumulation of fatigue loading and the concrete strength given that sufficient strengthening is secured.

- The insignificance of the change in the performance provided by the specimens with developed lengths of 67%, 80% and 93% of the CFRP tendon demonstrated that the developed length of the tendon has practically no effect on the strengthening performance given that appropriate tensioning has been secured. However, an appropriate and sufficient developed length should be secured to prevent the anchor be installed outside the effective depth d in which case shear failure would occur due to the increase of inclined cracks.

- The analysis of the ductility with respect to the concrete strength and the developed length of the tendon provided results complying with those of previous research. All the specimens exhibited nearly the same ductility according to the concrete strength and the ductility appeared to increase with longer developed length of the tendon. Accordingly, adopting longer developed length of the tendon was recommended for securing stable ductile behavior.

Author Contributions: Conceptualization, J.-S.P.; Methodology, H.B.P.; Validation, J.-Y.K., W.-T.J. and J.-S.P.; Analysis, H.B.P. and J.-S.P.; Investigation, J.-Y.K.; Writing—original draft preparation, H.B.P.; Writing—review and editing, H.B.P. and W.-T.J.; Supervision, H.B.P.; Project administration, W.-T.J.; Funding acquisition, J.-S.P.

Funding: This research was funded by a grant (17SCIP-B128496-01) from Smart Civil Infrastructure Research Program funded by Ministry of Land, Infrastructure and Transport of Korean government.

Conflicts of Interest: The authors declare no conflict of interest.

References

1. De-Lorenzis, L.; Micelli, F.; La Tagola, A. Passive and active near-surface mounted FRP rods for flexural strengthening of RC concrete beams. In Proceedings of the 3rd International Conference on Composites in Infrastructure, San Francisco, CA, USA, 10–12 June 2002.
2. El-Hacha, R.; Soudki, K. Prestressed near-surface mounted fibre reinforced polymer reinforcement for concrete structures—A review. *Can. J. Civ. Eng.* **2013**, *40*, 1127–1139. [CrossRef]
3. KICT. *Development of Bridge Strengthening Methods Using Prestressed FRP Composites*; Technical Report; KICT: Goyang-Si, Korea, 2015.
4. Liu, X.; Huang, L.; Xu, M.; Zhang, Z. Influence of different modeling strategies for CFRP on finite element simulation results. In Proceedings of the International Symposium on Materials Application and Engineering (SMAE 2016), MATEC Web of Conferences, Chiang Mai, Thailand, 20–21 August 2016.
5. Moon, J.-B.; Jang, H.-K.; Kim, C.-G. High velocity impact characteristics of MWNT added CFRP at LEO space environment. *Adv. Compos. Mater.* **2017**, *26*, 391–406. [CrossRef]
6. Ryan, S.; Schaefer, F.; Riedel, W. Numerical simulation of hypervelocity impact on CFRP/Al HC SP spacecraft structures causing penetration and fragment ejection. *Int. J. Impact Eng.* **2006**, *33*, 703–712. [CrossRef]
7. Xu, J.; El Mansori, M. Cutting modeling of hybrid CFRP/Ti composite with induced damage analysis. *Materials* **2016**, *9*, 22. [CrossRef] [PubMed]
8. Du, Z.; Zhu, M.; Wang, Z.; Yang, J. Design and application of composite platform with extreme low thermal deformation for satellite. *Compos. Struct.* **2016**, *152*, 693–703.
9. Al-Saadi, N.T.K.; Mohammed, A.; Al-Mahaidi, R. Fatigue performance of near-surface mounted CFRP strips embedded in concrete girders using cementitious adhesive made with graphene oxide. *Constr. Build. Mater.* **2017**, *148*, 632–647. [CrossRef]
10. Al-Saadi, N.T.K.; Mohammed, A.; Al-Mahaidi, R. Assessment of residual strength of concrete girders rehabilitated using NSM CFRP with cementitious adhesive made with graphene oxide after exposure to fatigue loading. *Constr. Build. Mater.* **2017**, *153*, 402–422. [CrossRef]
11. Quattlebaum, J.B.; Harries, K.A.; Petrou, M.F. Comparison of three flexural retrofit systems under monotonic and fatigue loads. *J. Bridge Eng.* **2005**, *10*, 731–740. [CrossRef]
12. Aidoo, J.; Harries, K.A.; Petrou, M.F. Full-scale experimental investigation of repair of reinforced concrete interstate bridge using CFRP materials. *J. Bridge Eng.* **2006**, *11*, 350–358. [CrossRef]
13. Rosenboom, O.; Rizkalla, S. Behavior of prestressed concrete strengthened with various CFRP systems subjected to fatigue loading. *J. Compos. Constr.* **2006**, *10*, 492–502. [CrossRef]
14. Badawi, M.; Soudki, K. Fatigue behavior of RC beams strengthened with NSM CFRP rods. *J. Compos. Constr.* **2009**, *13*, 415–421. [CrossRef]
15. Wahab, N.; Soudki, K.A.; Topper, T. Mechanics of bond fatigue behavior of concrete beams strengthened with NSM CFRP rods. *J. Compos. Constr.* **2011**, *15*, 934–942. [CrossRef]
16. ACI Committee 440. *Guide for the Design and Construction of Externally Bonded FRP Systems for Strengthening Concrete Structures*; ACI 440.2R-08; American Concrete Institute: Farmington Hills, MI, USA, 2008.
17. ACI Committee 215. *Considerations for Design of Concrete Structures Subjected to Fatigue Loading*; ACI 215R-74 (Reapproved 1997); American Concrete Institute: Farmington Hills, MI, USA, 2008.

18. *AASHTO LRFD Bridge Design Specifications*, 3rd ed.; American Association of State Highway and Transportation Officials: Washington, DC, USA, 2004.

19. *CSA Canadian Highway Bridge Design Code (CHBDC)*; CSA S6-06; CAS International: Toronto, ON, Canada, 2006.

20. Oudah, F.; El-Hacha, R. Fatigue behavior of RC beams strengthened with prestressed NSM CFRP rods. *Compos. Struct.* **2012**, *94*, 1333–1342. [CrossRef]

21. Brena, S.F.; Benouaich, M.A.; Kreger, M.E.; Wood, S.L. Fatigue tests of reinforced concrete beams strengthened using carbon fiber-reinforced polymer composites. *ACI Struct. J.* **2005**, *102*, 305–313.

22. Balaguru, P.; Shah, S.P. A method of predicting crack widths and deflections for fatigue loading. *ACI Spec. Publ.* **1982**, *75*, 153–176.

23. Jung, W.T.; Park, J.S.; Kang, J.Y.; Park, H.B. Strengthening Effect of Prestressed Near-Surface-Mounted CFRP Tendon on Reinforced Concrete Beam. *Adv. Mater. Sci. Eng.* **2018**, *2018*, 9210827. [CrossRef]

24. Oudah, F.; El-Hacha, R. Performance of RC beams strengthened using prestressed NSM-CFRP strips subjected to fatigue loading. *J. Compos. Constr.* **2012**, *16*, 300–307. [CrossRef]

25. KICT. *Development and Application of FRP Prestressing Tendon and Anchorage for Concrete Structures*; Construction & Transportation R&D Report; KAIA: Anyang-si, Korea, 2008.

materials

MDPI

Article

Influence of Abrasive Waterjet Parameters on the Cutting and Drilling of CFRP/UNS A97075 and UNS A97075/CFRP Stacks

Raul Ruiz-Garcia, Pedro F. Mayuet Ares *, Juan Manuel Vazquez-Martinez and Jorge Salguero Gómez

Department of Mechanical Engineering & Industrial Design, Faculty of Engineering, University of Cadiz, Av. Universidad de Cadiz 10, E-11519 Puerto Real-Cadiz, Spain; raul.ruizgarcia@uca.es (R.R.-G.); juanmanuel.vazquez@uca.es (J.M.V.-M.); jorge.salguero@uca.es (J.S.G.)
* Correspondence: pedro.mayuet@uca.es; Tel.: +34-616-852-858

Received: 30 November 2018; Accepted: 24 December 2018; Published: 30 December 2018

Abstract: The incorporation of plastic matrix composite materials into structural elements of the aeronautical industry requires contour machining and drilling processes along with metallic materials prior to final assembly operations. These operations are usually performed using conventional techniques, but they present problems derived from the nature of each material that avoid implementing One Shot Drilling strategies that work separately. In this work, the study focuses on the evaluation of the feasibility of Abrasive Waterjet Machining (AWJM) as a substitute for conventional drilling for stacks formed of Carbon Fiber Reinforced Plastic (CFRP) and aluminum alloy UNS A97050 through the study of the influence of abrasive mass flow rate, traverse feed rate and water pressure in straight cuts and drills. For the evaluation of the straight cuts, Stereoscopic Optical Microscopy (SOM) and Scanning Electron Microscopy (SEM) techniques were used. In addition, the kerf taper through the proposal of a new method and the surface quality in different cutting regions were evaluated. For the study of holes, the macrogeometric deviations of roundness, cylindricity and straightness were evaluated. Thus, this experimental procedure reveals the conditions that minimize deviations, defects, and damage in straight cuts and holes obtained by AWJM.

Keywords: AWJM; stack; CFRP; aluminum UNS A97050; SOM/SEM; kerf taper; surface quality; macrogeometric deviations

---

1. Introduction

Over the last few decades, the aeronautical industry has been highlighted for its capacity to develop and manufacture structural elements built with advanced materials, having achieved a leading position in this area of activity with respect to other sectors.

In this sense, the aeronautical industry has demonstrated its capacity for the development and manufacture of complex elements built with advanced materials. Thus, the main manufacturers (Airbus and Boeing) have increased the use of new materials, mainly plastic matrix composites, in combination with those traditionally used, such as Duralumin alloys of 2XXX or the Al-Zn of 7XXX series, with the aim of reducing aircraft weight, maintaining the structural integrity of the assembly. These materials have undoubted advantages linked to the demand of greater safety, and lower energy consumption and maintenance costs that characterize the air-transport today. Additionally, they provide an excellent relationship between mechanical strength and weight, rigidity and an increase in the life-cycle thanks their good behavior against fatigue and corrosion [1,2].

Most of the structural elements used in aircraft construction need to undergo different machining operations, mainly drilling or milling of contours, prior to assembly work through rivets in the Final

Assembly Lines (FAL) [3,4]. During the assembly tasks in aeronautical structures, these materials are joined in the form of stacks, which must be processed with drilling cycles under strict dimensional and geometric requirements, making it difficult to keep these tolerances under control when the nature of the materials is different [5–8].

Indeed, the combination of materials of a different nature has a negative impact during machining operations. On the one hand, both the heterogeneity of the material and the abrasive behavior of the carbon fiber negatively affect the tool life. Therefore, machining conditions and tool geometry must be adapted to these materials in order to reduce tool wear and thermal and mechanical defects produced during the cutting process, such as delamination or thermal damage to the composite matrix [9–11]. Moreover, Sorrentino et al. [12] demonstrated that Abrasive Waterjet Machining (AWJM) extends the high cycle fatigue strength of bolt holes and the fatigue life of bolted composite joints. On the other hand, aluminum alloys tend to modify the geometry of the tool [13], especially by the development of adhesive phenomena such as Build Up Layer (BUL) or Build Up Edge (BUE) [5,14]. The union of these phenomena causes accelerated wear of the tool through the loss of geometry and the increase in temperature reached during the cutting process, which causes a reduction in tool life due to the synergy of the wear mechanisms produced.

This is compounded by problems at the stack interface, such as burring and cleaning due to accumulated chip residues. As a result, the drilling process is complex to carry out in a single step [15]. Instead, different successive drilling steps must be carried out until the final diameter is obtained, including cleaning the rework at the interface, which does not allow One Way Assembly (OWA) to be achieved as a key technology for process automation.

Alternatively, some authors have conducted studies of machining stacks with unconventional technologies such as laser or AWJM [10,16–19]. In particular, AWJM has been widely studied as one of these machining alternatives to replace contour milling processes, although experimental studies are also beginning to appear, analyzing the influence of drilling on different aeronautical materials, Table 1. This is mainly due to different factors that positively affect the surface integrity of the final parts. Among them, and in comparison with conventional machining processes, the absence of tool wear, the reduction of residual stresses induced on the surface of the material and the reduction of surface thermal damage as a result of low cutting temperatures should be highlighted [10,20,21].

However, the AWJM process shows its own limitations that lead to the appearance of specific defects during the cutting process (Figure 1). The most common defects in the process are the kerf taper, the Erosion Affected Zone (EAZ) and the formation of three possible different roughness zones along the machined surface [22]:

- Initial Damage Region (IDR). The area where the water jet hits on the material producing EAZ. The roughness in this region is high due to the abrasive particles impacting the material.
- Smooth Cutting Region (SCR). The region of variable thickness depending on the cutting parameters. It is the region with the best surface quality because it does not suffer the impact of particles and the jet still has enough kinetic energy to cut.
- Rough Cutting Region (RCR). The final region where the jet ends of cut material. The jet has lost enough cutting capacity and produces macrogeometrical defects as striation marks.

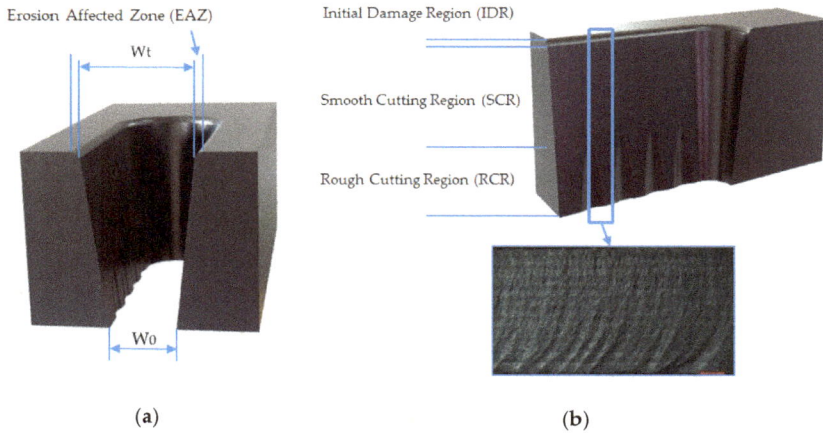

Figure 1. Scheme with the main defects associated with Abrasive Waterjet Machining (AWJM): (a) Erosion affected zone and kerf taper defined by inlet width (Wt) and outlet width (Wb); (b) different roughness zones that can be formed in AWJM.

Table 1. Comparative table with AWJM experimental studies on aeronautical materials.

Material	Thickness	Experimental	Parameters	Main Finding	Authors
CFRP/Ti-6Al-4V	10/11 mm	Straight Cuts	WP, TFR, AMFR	Taper analysis in stacks	Alberdi et al. [10]
CFRP	6 mm	Holes (6.35 mm)	WP, AMFR, SoD	Reduction of delamination	Phapale et al. [23]
CFRP	1.2 mm	Straight Cuts	WP, TFR, AMFR, SoD	Defect analysis	Schwartzentruber et al. [24]
CFRP	1.2 mm	Piercings	WP, TFR, AMFR, SoD	Piercing formation and delamination analysis	Schwartzentruber et al. [25]
Al 7075	7 mm	Straight Cuts	WP, TFR, SoD	Surface roughness analysis	Ahmed et al. [26]
GFRP	3.5 mm	Straight Cuts	WP, TFR, AMFR, SoD	Surface roughness analysis	Ming Ming et al. [27]
CFRP	6/12 mm	Straight Cuts	WP, AMFR	Taper and surface roughness analysis	Alberdi et al. [20]
CFRP	10.4 mm	Straight Cuts	WP, TFR, SoD	Taper analysis	El-Hofy et al. [16]
Ti-6Al-4V	5 mm	Straight Cuts	WP, TFR, AMFR	Taper and surface roughness analysis	Gnanavelbabu et al. [28]
GFRP	4 mm	Holes (10 mm)	WP, AMFR, SoD	Surface roughness and MRR	Prasad et al. [29]

Specifically, the removal of material through AWJM is produced by erosion caused by particles that impact the material at high velocity and affect each material differently. In the case of carbon fiber reinforced with plastic matrix, the formation of the erosion process produces the breakage of the fibers and the degradation of the matrix. This prevents the layers of the material from remaining bonded causing the formation of initial cracks that result in delaminations when abrasive particles penetrate between the layers of the composite [30].

However, some characteristic defects in the final part may occur as a result of the effect of the combination of different parameters. In this article a study based on the influence of the main cutting parameters on AWJM is carried out in order to reduce the appearance of the defects mentioned in stacks formed by the aluminum alloy UNS A97050 and Carbon Fiber Reinforced Plastic (CFRP). To this end, two experiments were carried out based on the operations most required in the machining of aeronautical structures: Straight cuts to analyze the cutting profile and drills to study the viability of the process. Finally, the state of the cuts was evaluated through the use of microscopic inspection techniques and macro and microgeometric deviations.

2. Materials and Methods

For the experimental development a CFRP AIMS 05-01-002 composite plate, Table 2, and a UNS A97075 aluminum alloy plate with a tensile strength of 496 MPa and a shear strength of 290 MPa have

been used. Both 5 mm thickness plates have been mechanically joined by eight bolts to obtain two
stack configurations: CFRP/UNS A97075 and UNS A97075/CFRP.

Table 2. Carbon Fiber Reinforced Plastic (CFRP) pieces features.

Type of Material	Composition	Production Method	Technical Specification
Layers of unidirectional carbon fiber with epoxy resin matrix and a symmetrical stacking sequence of (0/90/45/-45/45/-45)	Intermediate module fiber (66%) and epoxy resin (34%)	Pre-preg and autoclaved at 458° ± 5° at a pressure of 0.69 MPa	AIMS-05-01-002

As technological parameters, combinations were made for each configuration of the three most
significant parameters: Water pressure (WP), abrasive mass flow rate (AMFR) and traverse feed rating
(TFR), due to the influence analyzed in [31]. The separation distance was kept constant at 3 mm
throughout the experimental phase and the abrasive selected was garnet with an average particle size
of 80 μm in order to optimize aluminum penetration [32]. Under these considerations, the experimental
design based on levels shown in Table 3 was established.

Table 3. Parameters used for each configuration.

Test	WP (bar)	TFR (mm/min)	AMFR (g/min)
1	2500	15	170
2	2500	15	340
3	2500	30	170
4	2500	30	340
5	2500	45	170
6	2500	45	340
7	1200	15	170
8	1200	15	340
9	1200	30	170
10	1200	30	340
11	1200	45	170
12	1200	45	340

To carry out the tests, two experimental blocks for each stack were made. On the one hand, straight
cuts were made in order to study the influence on the kerf taper and the different roughness zones.
On the other hand, 8 mm holes were drilled to study macrogeometry due to the fact that 7.92 mm is a
common drill diameter used in the aeronautical industry. For this purpose, the experimental design and
pre-simulation were carried out using the CAD/CAM software Lantek® edition 34.02.02.02.02.02.02,
making a total of 48 tests mechanized with a TCI water jet cutting machine model BPC 3020.

For the evaluation of straight cuts, on the one hand, optical evaluation of the machined material
has been used by means of Stereoscopic Optical Microscopy (SOM) and Scanning Electronic Microscopy
(SEM) techniques, and on the other hand, electron dispersive spectroscopy (EDS) was used to analyze
the compositional state of the samples. A Nikon SMZ 800 stereo optical microscope was used
for the SOM inspection and the Hitachi SU 1510 microscope was used for the SEM inspection.
These techniques were used to study the incrustation of abrasive particles in the IDR zone and
in the delaminations produced. In addition, it was used to generate a deeper measurement of the kerf
taper. The literature tends to evaluate the taper as the difference between the cutting width of the
water inlet and the cutting width of the water outlet depending on the thickness of the plate [19,20,33]
as shown in Figure 2b. However, this process concurs in a high variability depending on two width
measures (W_{top} and W_{bottom}). Since the IDR may interfere with that extent, this paper proposes a new
methodology based on image processing methods, for which ImageJ and Microsoft Excel® software
were used. It consists of capturing the image of the cut and its subsequent digitalization in 10 points

with a non-linear distribution, as shown in Figure 2c. Then, a coefficient between W_{top} and W_{bottom} is usually obtained, however as it can be observed in Figure 2c, the representation of the cut would be unreal. Regarding Figure 2c, once you remove the IDR the shape of the cut is almost a vertical line. That is why, in this paper, measures to calculate the average width of the cut have been used.

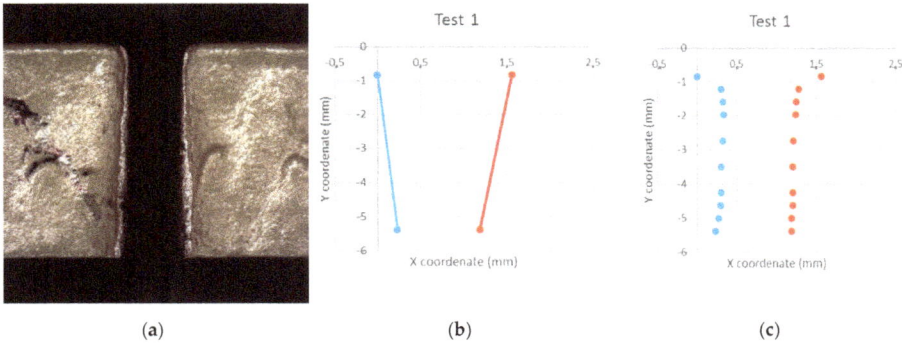

(a)   (b)   (c)

Figure 2. Proposal of measurement of the kerf taper from: (**a**) Stereoscopic Optical Microscopy (SOM) image; (**b**) geometry discretization; (**c**) traditional kerf assumption.

This new methodology does not provide a coefficient as taper measure, but an average distance. A distance that represents, in a more realistic way, the profile of the cut. However, in order to obtain an accurate result, data from the IDR has to be rejected, as is already regarded by some authors [34]. That way, the cut depth was divided into 10 measures with cosinoidal distribution, ensuring more measures density near the top and the bottom. Then, the measures that maintained a height variation regarding the next measure have been disposed, that is the case of the upper measure and the second one in Figure 2c. Taking all other measures, an average is calculated and that is the kerf taper vale proposed.

For the evaluation of the holes, a station of measurement Mahr MMQ44 Form Tester (Mahr, Göttingen, Germany) was used to measure the roundness at the entrance and exit of the drill in each material, the cylindricity of the entire profile of the drill, and the straightness in four separate generatrices to 90° as seen in Figure 3a. To analyze the macrogeometric deviations, replicas of the holes due to the impossibility of direct measuring on the material were fabricated. These replicas were made with a polymer type F80 Ra (R.G.X, Plastiform, Madrid, Spain) with the ability to guarantee stability during the measurement process for diameters greater than 4 mm. It is a two component polymer that solidifies after mixture. Plastiform provides a tool that ensures correct mixture while the polymers are injected into the hole that was replicated. It is a manual process that leads to hole replicas after 10 min of polymer solidification.

For the measurement of roughness, the Mahr Perthometer Concept PGK 120 (Mahr, Göttingen, Germany) was employed, as shown in Figure 3. This measurement was focused on the parameter Average Roughness (Ra), since it is one of the most used roughness parameters in the literature.

Ra analysis performed to the specimens in each test was carried out in three different zones coinciding with IDR, SCR and RCR (Figure 4). That way, six measures were obtained for each test performed, making a total of 144 roughness measurements.

Finally, to distinguish the most significant parameters for evaluation results, analysis of variance (ANOVA) for a 95% confidence interval was employed. After that, contour charts for each variable studied in the experimental were obtained.

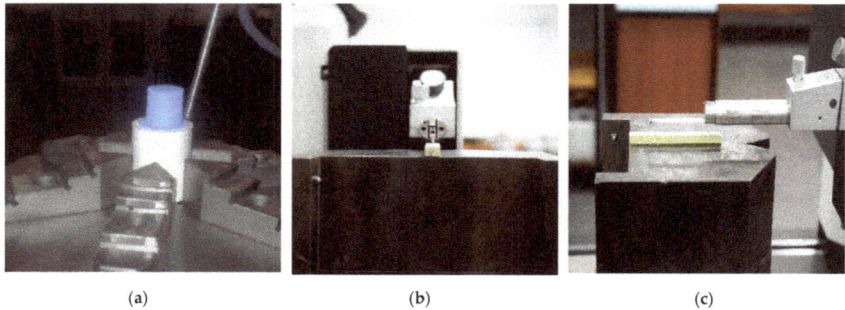

Figure 3. (**a**) Measure of replica to obtain geometrical results; (**b**) measure of roughness of UNS A97075, front view; (**c**) measure of roughness of UNS A97075, side view.

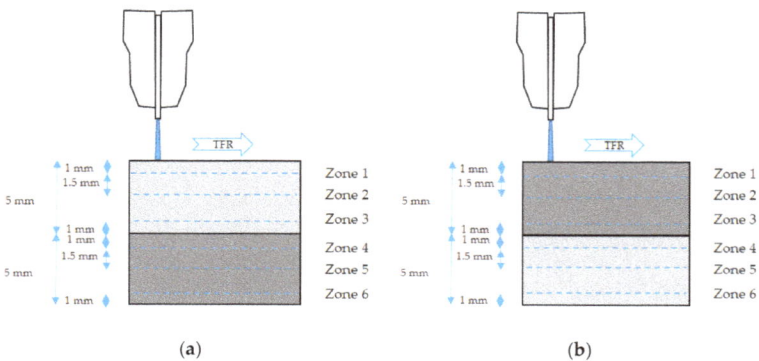

Figure 4. Schematic representing the roughness measurement zones and the distance between them for: (**a**) UNS A97075/CFRP configuration; (**b**) CFRP/UNS A97075 configuration.

3. Results and Discussion

3.1. Straight Cuts Evaluation

3.1.1. SOM/SEM Evaluation

SOM inspection was carried out of both jet entrances into the stack (Figure 5), and along the cut profile. This way, the jet variations contribution to the kerf profile can be observed, phenomenon related to damages produced in the IDR zone [31].

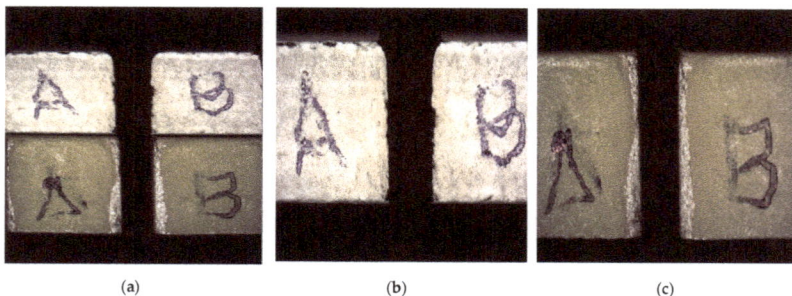

Figure 5. SOM image of the cutting front in: (**a**) Stack CFRP/UNS A97075; (**b**) CFRP plate; (**c**) UNS A97075 plate.

On the other hand, Figure 6 shows the profile of CFRP specimens in order to identify delaminations. In order to visualize the delamination along the machined surface, several images were taken showing the absence of visible delamination after machining in the test performed with the parameters considered to be the most aggressive.

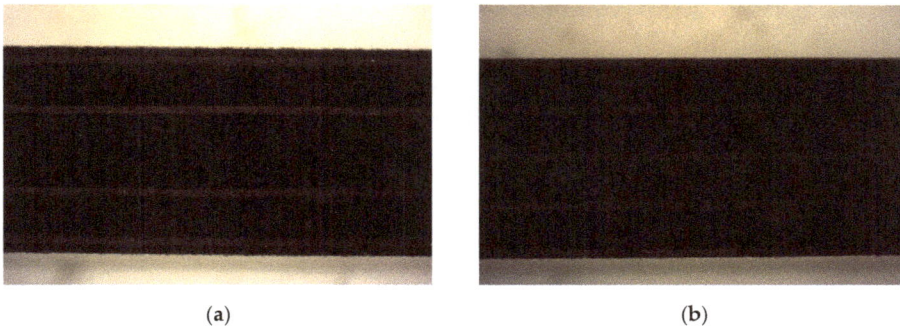

(a)   (b)

Figure 6. SOM image of CFRP profile. Test 2. Water pressure (WP) = 2500 bar, traverse feed rating (TFR) = 15 mm/min and abrasive mass flow rate (AMFR) = 340 gr/min for: (a) UNS A97075/CFRP; (b) CFRP/UNS A97075.

Figure 7 shows the results of the SEM inspection in CFRP showing that no delamination was detected. However, Figure 7c shows in detail the state of the specimen entrance zone where signs of impact deformation and particle drag were observed. This state extends to the interface reflecting that a percentage of particles have lodged in the space between the two materials.

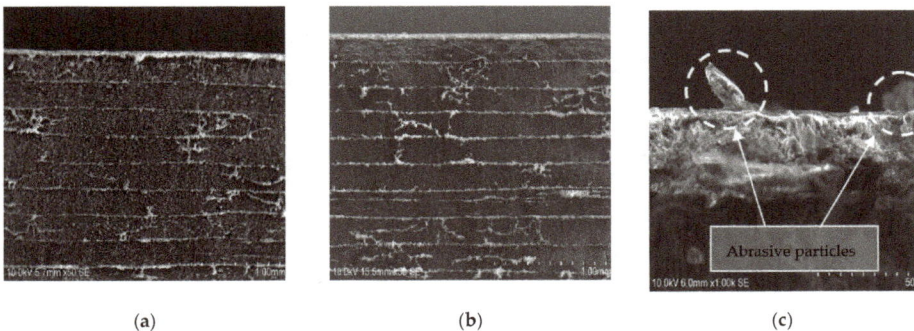

(a)   (b)   (c)

Figure 7. SEM image of CFRP profile. Test 2. WP = 2500 bar, TFR = 15 mm/min and AMFR = 340 gr/min for (a) UNS A97075/CFRP; (b) CFRP/UNS A97075; (c) Abrasive particles in the interface over CFRP.

As for aluminum alloy, SOM study showed a series of dark colored streaks along the profile that repeated for both configurations to a greater or lesser extent depending on the energy of the jet. Specifically, Figure 8a shows the marks mentioned at the bottom while Figure 8b at the top. This phenomenon, together with the color of the stretch mark, seems to indicate that they are located in the zone close to the contact with the carbon fiber. Finally, Figure 8c shows the result of the study for test 11 where no transfer of carbon fiber to aluminum is observed, possibly due to the lower WP and AMFR and thus, lesser jet kinetic energy resulting in an inferior material removal rate [10].

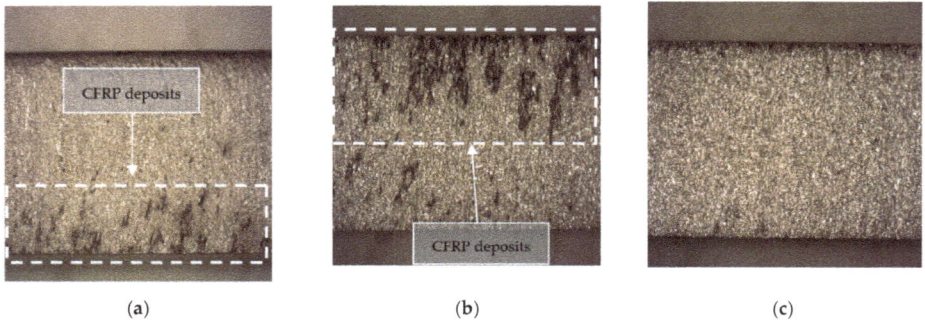

(a) (b) (c)

Figure 8. Profile SOM of UNS A97075 from: (**a**) UNS A97075/CFRP. Test 2. WP = 2500 bar, TFR = 15 mm/min and AMFR = 340 gr/min; (**b**) CFRP/UNS A97075. Test 2. WP = 2500 bar, TFR = 15 mm/min and AMFR = 340 gr/min; (**c**) CFRP/UNS A97075. Test 11. WP = 1200 bar, TFR = 45 mm/min and AMFR = 170 gr/min.

In an attempt to obtain more information on the marks observed in Figure 8, the SEM/EDS inspection of aluminum was focused on discovering the state of the aluminum and the nature of these marks. Initially, Figure 9a,b shows the state of the material at the inlet. In a detailed way, the embedded particles and the deformation produced during the cutting process are appreciated, coinciding with the IDR or zone 1.

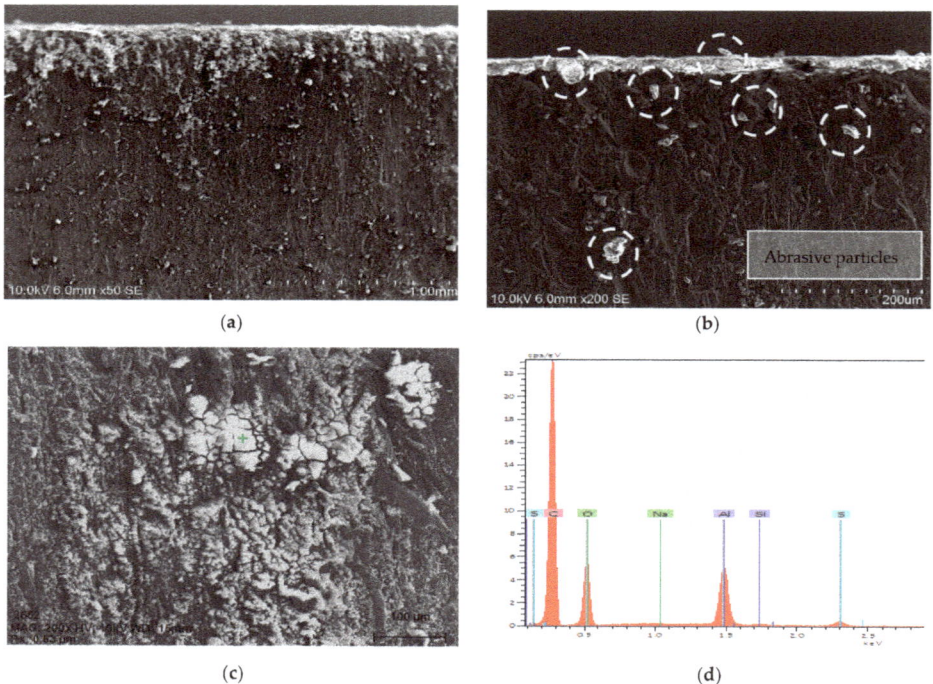

(a) (b)

(c) (d)

Figure 9. Test 2 Scanning Electron Microscopy (SEM) evaluation: (**a**) Abrasive imbued into UNS A97075; (**b**) abrasive particles in the interface and channel created over the material; (**c**) remains of carbon and point of EDS; (**d**) EDS results with a peak on the carbon.

On the other hand, Figure 9c,d shows the stain examined in the striations observed by SOM microscopy and the results of the EDS analysis, respectively. The EDS analysis revealed the high presence of carbon at this point, confirming the carry-over of carbon particles during machining from one material to another. It should be noted that no traces of aluminum deposited on the carbon fiber were detected.

To analyze the state of the aluminum outside the zone of the stretch marks, another EDS spot was carried out outside those stains and showed almost no carbon and a huge peak on aluminum. As a direct conclusion, it appears that particles from composite are swept for the water beam and because of the water high energy, they end up embedded into UNS A97075. It seems like composite deposition over aluminum has a direct correlation with beam penetration capacity.

Therefore, contrary to what one would expect, a higher abrasive pressure and flow has not resulted in an increase in delaminations for both configurations. Similarly, the inclusion of abrasive particles has not greatly increased within the parameters studied. However, an increase in the inclusion of carbon particles in the aluminum alloy was observed as the pressure increases.

3.1.2. Kerf Taper Evaluation

The ANOVA analysis performed showed that AMFR and TFR parameters were the most influential in taper formation. Average kerf taper values for each material when the configuration UNS A97075/CFRP is set are shown in Figure 10. The same values for configuration CFRP/UNS A97075 are shown in Figure 11. For deeper research, Appendix A shows the average taper for each test and its standard deviation.

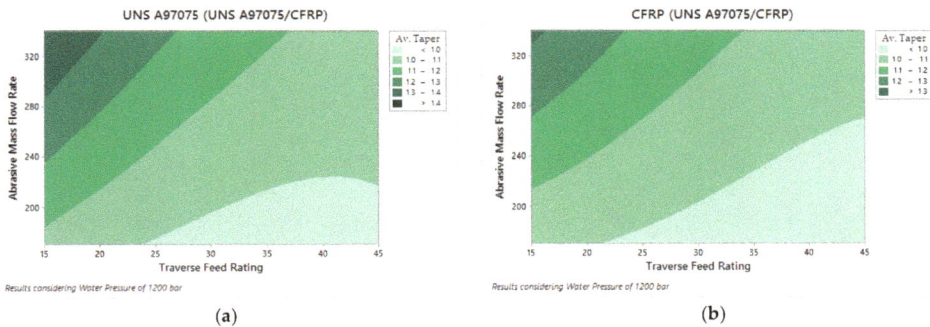

Figure 10. Average kerf taper for USN A97075/CFRP configuration: (a) UNS A97075; (b) CFRP.

Figure 10 shows a wide parameters combination that maintain kerf taper below 1.0 mm. This means a wide parameters combination that minimizes material removal percentage and leaves a more precise cut.

In this way, the data represented in Figures 10 and 11 show that the taper is reduced as AMFR decreases and TFR increases, showing the best results for TFR = 45 mm/min and AMFR = 170 gr/min, in accordance with [33].

This behavior is shared with the CFRP behavior in CFRP/UNS A97075 configuration. Figure 11b, however, shows a very different behavior. This change is due to the lesser energy of the water beam when it collides with the aluminum. Since a percentage of energy is transformed during the CFRP machining, it appears that the AMFR is the determinant parameter when the cut's width is examined. As for the differences between the two material configurations, Figure 11 shows that when the jet directly affects the carbon fiber, the taper generated for the best parameter ratio reaches values higher than 1.2 compared to the value 1 reached for the UNS A97075/CFRP configuration. This shows the difference in the mechanical properties of each material, offering greater resistance to penetration of the metallic material.

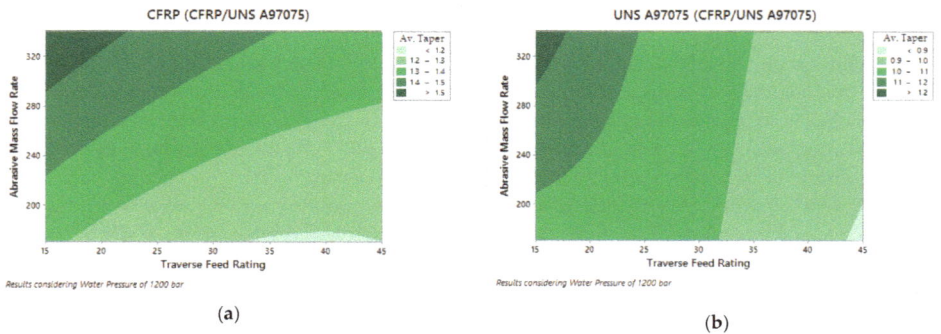

(a)

(b)

Figure 11. Average kerf taper for CFRP/USN A97075/ configuration: (**a**) CFRP; (**b**) UNS A97075.

Overall, a similar behavior is observed between the materials located in the upper and lower part of the stack. Despite this, a smaller taper is always observed in UNS A97075 than in CFRP.

3.1.3. Surface Roughness

The influential parameters in the analysis of surface quality are also AMFR and TFR for both configurations.

Figure 12 shows the results of the UNS A97075/CFRP configuration. A tendency to increase the roughness can be observed as TFR increases and AMFR decreases. Figure 13, on the other hand, shows the results of CFRP/UNS A97075 configuration. The same trend as in Figure 12 is observed although exist difference between the material placed at the top and bottom.

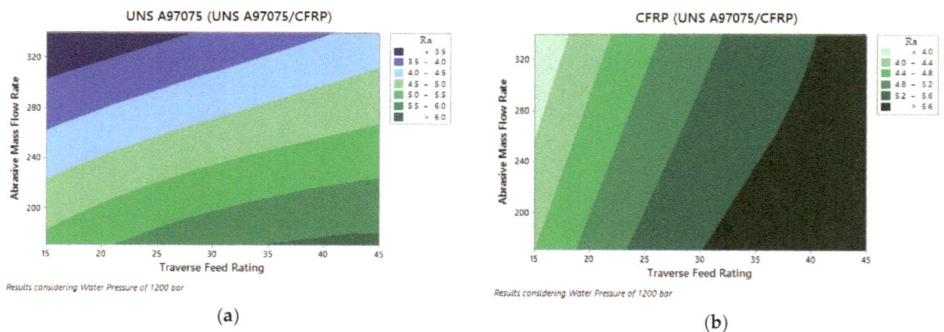

(a)

(b)

Figure 12. Roughness for UNS A97075/CFRP configuration: (**a**) UNS A97075 and (**b**) CFRP.

The data show that the AMFR parameter has a greater influence on aluminum, especially when it is at the exit of the material. This effect can be seen in the horizontality of the contour graph studied (Figure 13b). As for the composite material, it presents influence of TFR and AMFR for both configurations. In this sense, the data show that the composite material has slightly lower values than the metal alloy because the use of low pressures favors a better surface quality in CFRP to oppose less resistance to cutting. On the other hand, this means that the aluminum registers higher roughness data due to the low kinetic energy of the jet, favoring the appearance of defects in the different areas studied.

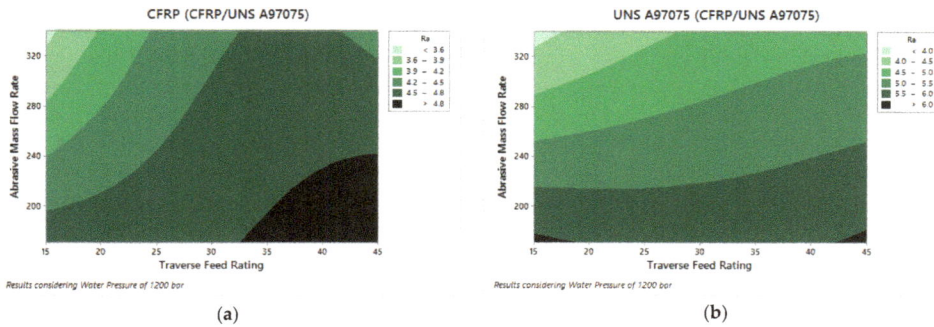

Figure 13. Roughness for CFRP/USN A97075 configuration: (**a**) CFRP and (**b**) UNS A97075.

A more in-depth analysis of the data based on Appendix B reflects that the area with the greatest damage is region 1 or IDR due to deformations and damage caused by the impact of the jet on the material. In addition, this is the region where embedded particles were detected. On the other hand, the material at the bottom has lower roughness values in region 4 due to the protection of the material at the top.

On the other hand, it can be observed that regions 2 and 5, corresponding to SCR, do not have values lower than those recorded in zones 3 and 6 as RCR. This indicates the existence of two zones because the jet still has enough kinetic energy to make the cut without the appearance of striations.

3.2. Holes Evaluation

3.2.1. Roundness Deviation

Figure 14 shows the data obtained from roundness deviations for each material and the total average of both materials. Readers can also find Appendix C with measured data and its standard deviation. In this way, the results can be analyzed separately.

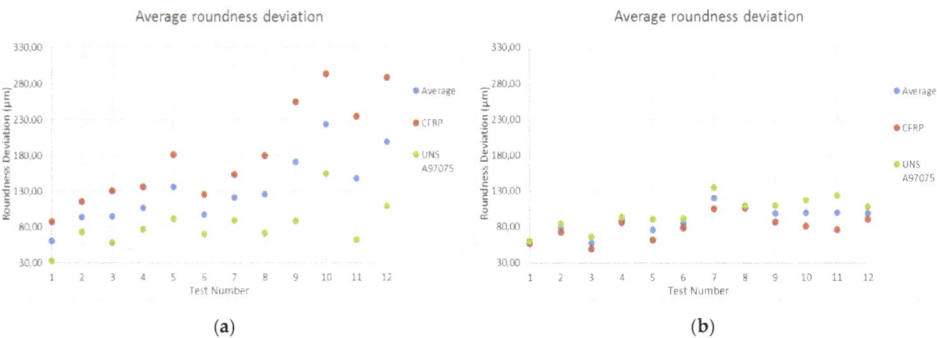

Figure 14. Average roundness: (**a**) UNS A97075/CFRP configuration; (**b**) CFRP/UNS A97075 configuration.

Figure 14a shows the data for UNS A97075/CFRP configuration. The data show that in all tests the deviation is higher for CFRP, even though it is the material located at the bottom of the stack. This is due to the fact that the erosion and removal of composite materials is different from that produced in metallic materials. Thus, in CFRP the particles weaken and remove the matrix of the compound to subsequently break the fibers of the adjacent zone and in Al the process of material elimination takes place due to the micromachining produced by the edges of the abrasive particles, being more homogeneous the elimination of material in metallic materials. [2]. This phenomenon, combined with the material's resistance to jet dispersion as a result of the loss of energy after cutting the aluminum, leads to an increase in the deflection in this material. This deflection increases considerably as WP decreases and TFR increases, which makes sense because these are tests with lower shear power.

On the other hand, Figure 14b shows the results of CFRP/UNS configuration A97075. In this particular case, the deviations follow a similar relationship to that of the previous case in terms of parameter influence, although it is true that the difference in the measured values is high. Thus, although in this case the aluminum is at the bottom of the pile, it seems that the expansion of the water jet does not deform the entry zone due to the differences in terms of removal of material explained in the previous paragraph. This results in homogeneous deviations in roundness between the two configurations, which favors a subsequent joint by means of rivets.

3.2.2. Cylindricity Deviation

Cylindricity deviation was also measured with two measures for each material and configuration. However, due to the nature of the test, only one value results as output. Appendix D contains all collected data. Nevertheless, an ANOVA description of variables influence is shown in Figure 15.

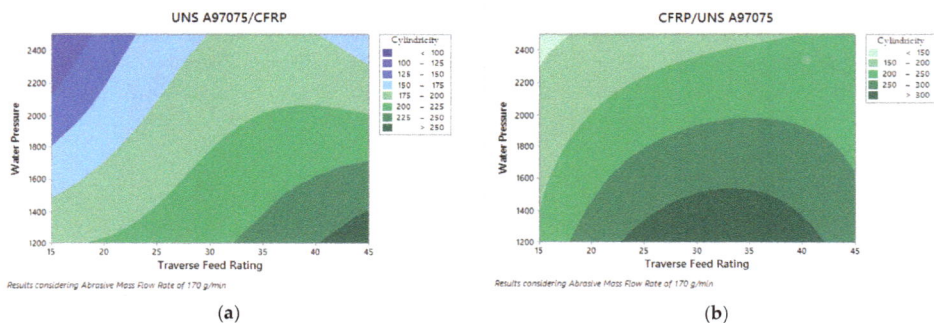

Figure 15. Cylindricity results on: (a) UNS A97075/CFRP configuration; (b) CFRP/UNS A97075 configuration.

The ANOVA analysis carried out shows that the parameters that have the greatest influence on the formation of the deviations are WP and TFR. Specifically, Figure 15 shows that the UNS configuration A97075/CFRP has lower cylindricity deviation values, which is in good agreement with the taper values obtained. This is due to the close relationship between both parameters. In order to offer a better correlation of results, the profiles measured for test 11 are presented as an example (Figure 16).

A more detailed description of the data reflects that cylindricity decreases as TFR decreases and WP increases. Specifically, Figure 15a reveals that TFR has a superior influence when the alloy is at the inlet of the material which reflects the importance of employing reduced feed rates to prevent its formation. As for Figure 16, both (a) and (b) show CFRP on the bottom and UNS A97075 on the top of the cone. It can be observed how it affects the energy loss to the generated hole, especially in Figure 16a.

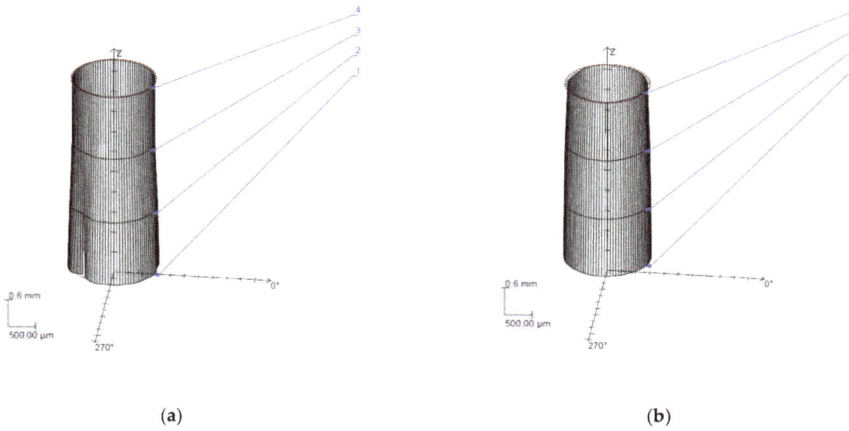

(a)   (b)

Figure 16. Cylindricity deviations. Test 11. WP = 1200 bar, TFR = 45 mm/min and AMFR = 170 gr/min for: (**a**) UNS A97075/CFRP configuration; (**b**) CFRP/UNS A97075 configuration.

3.2.3. Straightness Deviation

In this case, there is no distinction between materials and straightness was evaluated throughout the entire profile. Thus, Figure 17 shows a comparison between the values obtained for the two configurations. However, Appendix E shows all measured data with the numerical value of the standard deviation.

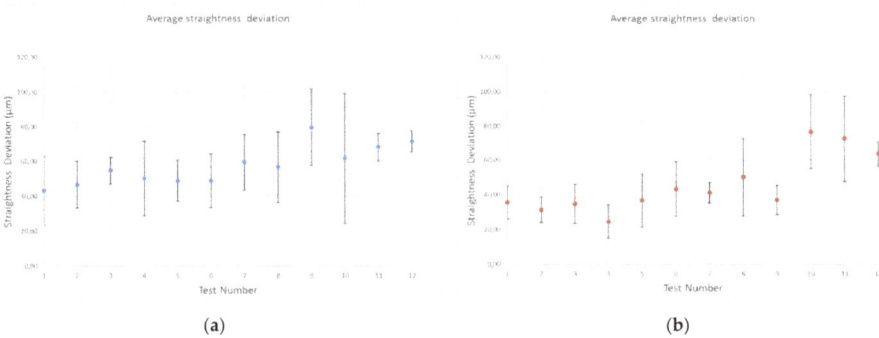

(a)   (b)

Figure 17. Straightness deviations results on: (**a**) UNS A97075/CFRP configuration; (**b**) CFRP/UNS A97075 configuration.

As a general conclusion, a higher water jet drilling capacity means less straightness deviation. It is also observed that the CFRP/UNS A97075 configuration shows better results for the same test number except for the last three tests which, due to their lower drilling capacity due to the use of lower WP and TFR, are not able to maintain a uniform cutting profile of the aluminum alloy and therefore cannot maintain straightness along the hole.

The results reveal that the data in configuration UNS A97075/CFRP are slightly lower than those recorded in configuration CFRP/UNS A97075. In addition, it should be noted that for high pressures the straightness deviation increases when the compound is located at the top.

On the other hand, it should be noted that the standard deviation presented by the results is high, which makes it difficult to establish relationships between the results. This major standard deviation is directly related to the greater cylindricity deviation of some tests.

4. Conclusions

A study has been carried out on the influence of the parameters of the abrasive water jet on the quality of straight cuts and holes in composite materials and aeronautical aluminum. Based on this, the following conclusions can be drawn:

1. The machining of straight cuts has revealed that thermal damage is eliminated and the appearance of delamination in CFRP is reduced. Thus, for the selected parameters, no delamination was found in the mechanized test samples.
2. The proposed kerf taper measurement method was validated for measurement in stacks. The results show the influence of the selected parameters obtaining the best results for high TFR and AMFR for both configurations, especially USN A97075/CFRP, with CFRP being the material with the highest kerf taper. On the other hand, the CFRP/UNS A97075 configuration has lower microgeometric deviations for the three evaluated parameters due to the lower loss of jet energy.
3. Ra is in all cases below 7 µm, although this value is specific for tests 9, 10, 11 and 12. The functional holes show a lower roughness for both materials in any configuration. Nevertheless, it appears that the UNS A97075/CFRP configuration offers a better roughness of the holes.
4. The study of surface quality has revealed that the IDR zone of the second material (region 4) is attenuated from impacts of particle and EAZ impacts. On the other hand, the presence of RCR was not detected.
5. The measurements obtained of roundness present a greater deviation at the entrance of the drill due to the IDR zone in region 1, independently of the selected configuration, although it is true that the CFRP/UNS A97050 configuration presents values around 200% lower for the tests with lower penetration power (9, 10, 11 and 12).
6. The influence of kerf taper on cylindricity deviations was reflected through the evaluated profiles, recording that the parameters with the greatest influence on its formation are WP and TFR. In this case, the configuration UNS A97075/CFRP presents better results of cylindricity.
7. The straightness deviations did not allow consolidated conclusions to be drawn due to the high standard deviationHowever, it can be seen once again that tests 9, 10, 11 and 12 have higher values.

Finally, it should be noted that this process does not generate burrs in metallic materials due to its abrasive nature or thermal gradients that damage the material. On the other hand, it should be noted that each configuration has different characteristics, but it is the UNS A97075/CFRP configuration that presents the best results in terms of macro and microgeometric deviations.

Author Contributions: Conceptualization, R.R.-G. and P.F.M.A.; Methodology, R.R.-G. and P.F.M.A.; Software, R.R.-G. and J.S.G; Validation, P.F.M.A. and J.M.V.-M.; Formal Analysis, P.F.M.A. and J.M.V.-M.; Investigation, R.R.-G. and P.F.M.A.; Resources, J.M.V.-M. and J.S.G.; Data Curation, R.R.-G. and J.M.V.-M.; Writing-Original Draft Preparation, R.R.-G.; Writing-Review & Editing, P.F.M.A. and J.S.G.; Visualization, P.F.M.A. and J.M.V.-M.; Supervision, P.F.M.A; Project Administration, J.S.G.; Funding Acquisition, J.S.G.

Funding: This work has received financial support from Special Actions for the Transfer of the Programme for the Promotion and Promotion of the Research and Transfer Activity of the University of Cadiz.

Conflicts of Interest: The authors declare no conflict of interest.

Appendix A

Table A1. Average kerf taper data. CFRP (CFRP/UNS A97075 conf.).

Test	1	2	3	4	5	6	7	8	9	10	11	12
Av. Kerf Taper (mm)	1.28	1.58	1.20	1.40	1.15	1.43	1.32	1.57	1.21	1.44	1.20	1.35
Standard Deviation	0.16	0.26	0.18	0.23	0.13	0.29	0.18	0.24	0.15	0.24	0.16	0.25

Table A2. Average kerf taper data. UNS A97075 (CFRP/UNS A97075 conf.).

Test	1	2	3	4	5	6	7	8	9	10	11	12
Av. Kerf Taper (mm)	1.09	1.25	1.01	1.12	1.00	1.07	1.06	1.25	1.01	1.04	0.88	0.97
Standard Deviation	0.04	0.04	0.04	0.05	0.09	0.03	0.06	0.16	0.16	0.20	0.08	0.15

Appendix B

Table A3. Average roughness UNS A97075/CFRP configuration.

		UNS A97075/CFRP Configuration Ra (µm)											
	Test	1	2	3	4	5	6	7	8	9	10	11	12
Al (inlet)	Zone 1	4.95	4.00	6.73	4.74	6.68	6.43	7.39	3.36	8.00	3.76	8.22	5.05
	Zone 2	3.52	2.78	4.34	4.01	4.66	3.84	4.19	2.88	5.05	3.23	5.45	3.82
	Zone 3	4.22	3.45	4.79	3.27	5.04	3.79	3.87	2.85	4.58	3.69	4.70	3.64
CFRP (Outlet)	Zone 4	4.99	3.72	5.77	4.42	5.48	4.73	4.36	3.30	5.44	4.24	4.49	4.72
	Zone 5	5.39	4.48	5.54	4.42	6.31	4.94	4.31	3.84	5.89	5.07	5.98	5.40
	Zone 6	4.19	3.99	5.27	4.43	4.20	4.27	4.49	3.77	5.45	5.90	6.53	7.00

Table A4. Average roughness CFRP /UNS A97075 configuration.

		CFRP/ UNS A97075 Configuration Ra (µm)											
	Test	1	2	3	4	5	6	7	8	9	10	11	12
CFRP (inlet)	Zone 1	6.99	4.79	7.80	5.66	7.46	6.52	8.36	4.01	7.89	5.08	7.59	5.97
	Zone 2	4.78	3.80	5.04	4.49	5.38	4.34	5.00	3.54	5.17	4.54	5.00	4.40
	Zone 3	4.96	3.91	4.97	4.25	5.13	4.78	4.97	3.94	4.56	4.10	5.64	4.27
Al (Outlet)	Zone 4	4.54	3.42	5.12	4.17	5.19	4.01	4.57	3.55	4.62	4.19	4.48	4.06
	Zone 5	4.68	3.21	4.82	3.81	5.21	4.08	4.27	3.35	4.88	4.58	5.69	4.73
	Zone 6	4.22	3.12	5.11	3.35	5.31	4.24	5.20	3.63	4.79	4.54	5.10	4.44

Appendix C

Table A5. Roundness deviation (µm) for UNS A97075/CFRP conf.

Test	Measure 1	Measure 2	Measure 3	Measure 4	Average UNS A97075	Average CFRP	Average Measure	Standard Deviation
1	96.48	77.45	31.88	33.66	59.87	86.97	32.77	32.25
2	104.99	124.57	60.73	84.91	93.80	114.78	72.82	27.35
3	146.17	114.25	60.50	55.01	93.98	130.21	57.76	43.87
4	151.99	120.05	89.54	62.76	106.09	136.02	76.15	38.53
5	203.90	158.20	93.42	88.54	136.02	181.05	90.98	55.28
6	142.07	107.67	74.69	64.09	97.13	124.87	69.39	35.24
7	168.89	138.31	96.59	81.08	121.22	153.60	88.84	39.93
8	210.01	149.35	84.31	58.12	125.45	179.68	71.22	68.18
9	315.68	193.28	100.37	74.27	170.90	254.48	87.32	109.20
10	319.79	266.90	181.39	126.88	223.74	293.35	154.14	86.15
11	275.57	192.82	78.70	43.97	147.77	234.20	61.34	106.31
12	329.71	246.98	147.87	69.13	198.42	288.35	108.50	113.82

Table A6. Roundness deviation (µm) for CFRP/UNS A97075 conf.

Test	Measure 1	Measure 2	Measure 3	Measure 4	Average UNS A97075	AverageCFRP	Average Measure	Standard Deviation
1	66.51	45.39	60.58	59.34	57.96	55.95	59.96	8.94
2	71.87	71.80	78.91	88.29	77.72	71.84	83.60	7.80
3	47.26	49.77	47.57	83.42	57.01	48.52	65.50	17.65
4	86.57	83.68	94.45	91.91	89.15	85.13	93.18	4.91
5	58.74	63.59	83.87	95.32	75.38	61.17	89.60	17.18
6	85.05	71.66	90.04	90.63	84.35	78.36	90.34	8.82
7	99.85	109.79	132.34	137.40	119.85	104.82	134.87	17.94
8	110.65	99.89	110.77	107.45	107.19	105.27	109.11	5.10
9	88.70	83.88	102.59	115.27	97.61	86.29	108.93	14.20
10	76.02	85.78	106.88	126.58	98.82	80.90	116.73	22.55
11	65.05	85.53	111.45	134.05	99.02	75.29	122.75	30.10
12	88.18	90.90	86.37	127.23	98.17	89.54	106.80	19.46

Appendix D

Table A7. Cylindricity deviation (μm) for UNS A97075/CFRP conf.

Test	1	2	3	4	5	6	7	8	9	10	11	12
Cylindricity	138.96	199.66	183.62	194.69	203.58	180.12	212.42	229.93	336.68	355.84	273.40	332.66

Table A8. Cylindricity deviation (μm) for CFRP/UNS A97075 conf.

Test	1	2	3	4	5	6	7	8	9	10	11	12
Cylindricity	95.57	155.5	175.33	172.08	159.12	179.38	197.43	271.28	219.75	270.84	267.94	293.31

Appendix E

Table A9. Straightness deviation (μm) for UNS A97075/CFRP conf.

Test	Measure 1	Measure 2	Measure 3	Measure 4	Average Measure	Standard Deviation
1	58.46	62.10	27.46	24.83	43.21	19.79
2	30.48	41.43	53.22	61.07	46.55	13.41
3	52.21	45.82	56.84	63.90	54.69	7.62
4	46.63	71.48	22.00	61.11	50.31	21.45
5	41.06	38.27	52.47	63.85	48.91	11.70
6	46.41	27.91	61.74	59.18	48.81	15.46
7	79.89	59.42	58.25	40.77	59.58	16.00
8	26.74	65.06	71.89	63.99	56.92	20.42
9	99.42	97.71	66.59	55.38	79.78	22.19
10	103.27	77.93	50.00	15.98	61.80	37.50
11	66.76	71.13	77.09	58.08	68.27	8.00
12	78.43	74.81	64.45	68.96	71.66	6.19

Table A10. Straightness deviation (μm) for CFRP/UNS A97075 conf.

Test	Measure 1	Measure 2	Measure 3	Measure 4	Average Measure	Standard Deviation
1	22.10	42.04	35.09	42.44	35.42	9.50
2	33.85	20.39	35.17	35.23	31.16	7.21
3	31.80	49.59	22.58	34.68	34.66	11.21
4	35.72	25.12	24.92	12.10	24.47	9.66
5	39.70	14.82	42.47	49.72	36.68	15.17
6	57.85	54.92	35.09	25.16	43.26	15.74
7	42.72	48.14	39.55	33.75	41.04	6.02
8	45.42	45.46	28.12	81.73	50.18	22.56
9	28.63	36.15	48.45	33.84	36.77	8.40
10	79.12	92.27	45.22	89.97	76.65	21.72
11	36.07	87.57	89.38	76.94	72.49	24.89
12	55.17	62.75	72.29	64.42	63.66	7.02

References

1. Bazli, M.; Ashrafi, H.; Jafari, A.; Zhao, X.L.; Gholipour, H.; Oskouei, A.V. Effect of thickness and reinforcement configuration on flexural and impact behaviour of GFRP laminates after exposure to elevated temperatures. *Compos. Part B Eng.* **2019**, *157*, 76–99. [CrossRef]
2. Hejjaji, A.; Zitoune, R.; Crouzeix, L.; Roux, S.L.; Collombet, F. Surface and machining induced damage characterization of abrasive water jet milled carbon/epoxy composite specimens and their impact on tensile behavior. *Wear* **2017**, *376–377*, 1356–1364. [CrossRef]
3. Casalegno, V.; Salvo, M.; Rizzo, S.; Goglio, L.; Damiano, O.; Ferraris, M. Joining of carbon fibre reinforced polymer to Al-Si alloy for space applications. *Int. J. Adhes. Adhes.* **2018**, *82*, 146–152. [CrossRef]

4. Lambiase, F.; Durante, M.; Ilio, A. Di Fast joining of aluminum sheets with Glass Fiber Reinforced Polymer (GFRP) by mechanical clinching. *J. Mater. Process. Technol.* **2016**, *236*, 241–251. [CrossRef]

5. Park, K.-H.; Beal, A.; Kim, D.; Kim, D.W.; Kwon, P.; Lantrip, J. Tool wear in drilling of composite/titanium stacks using carbide and polycrystalline diamond tools. *Wear* **2011**, *271*, 2826–2835. [CrossRef]

6. Ramulu, M.; Branson, T.; Kim, D. A study on the drilling of composite and titanium stacks. *Compos. Struct.* **2001**, *54*, 67–77. [CrossRef]

7. Zitoune, R.; Krishnaraj, V.; Collombet, F. Study of drilling of composite material and aluminium stack. *Compos. Struct.* **2010**, *92*, 1246–1255. [CrossRef]

8. Kuo, C.; Li, Z.; Wang, C. Multi-objective optimisation in vibration-assisted drilling of CFRP/Al stacks. *Compos. Struct.* **2017**, *173*, 196–209. [CrossRef]

9. Saleem, M.; Toubal, L.; Zitoune, R.; Bougherara, H. Investigating the effect of machining processes on the mechanical behavior of composite plates with circular holes. *Compos. Part A Appl. Sci. Manuf.* **2013**, *55*, 169–177. [CrossRef]

10. Alberdi, A.; Artaza, T.; Suárez, A.; Rivero, A.; Girot, F. An experimental study on abrasive waterjet cutting of CFRP/Ti6Al4V stacks for drilling operations. *Int. J. Adv. Manuf. Technol.* **2016**, *86*, 691–704. [CrossRef]

11. Sayuti, M.; Sarhan, A.A.D.; Salem, F. Novel uses of SiO$_2$ nano-lubrication system in hard turning process of hardened steel AISI4140 for less tool wear, surface roughness and oil consumption. *J. Clean. Prod.* **2014**, *67*, 265–276. [CrossRef]

12. Sorrentino, L.; Turchetta, S.; Bellini, C. A new method to reduce delaminations during drilling of FRP laminates by feed rate control. *Compos. Struct.* **2018**, *186*, 154–164. [CrossRef]

13. D'Orazio, A.; El Mehtedi, M.; Forcellese, A.; Nardinocchi, A.; Simoncini, M. Tool wear and hole quality in drilling of CFRP/AA7075 stacks with DLC and nanocomposite TiAlN coated tools. *J. Manuf. Process.* **2017**, *30*, 582–592. [CrossRef]

14. Zitoune, R.; Krishnaraj, V.; Sofiane Almabouacif, B.; Collombet, F.; Sima, M.; Jolin, A. Influence of machining parameters and new nano-coated tool on drilling performance of CFRP/Aluminium sandwich. *Compos. Part B Eng.* **2012**, *43*, 1480–1488. [CrossRef]

15. Wang, F.; Qian, B.; Jia, Z.; Fu, R.; Cheng, D. Secondary cutting edge wear of one-shot drill bit in drilling CFRP and its impact on hole quality. *Compos. Struct.* **2017**, *178*, 341–352. [CrossRef]

16. El-Hofy, M.; Helmy, M.O.; Escobar-Palafox, G.; Kerrigan, K.; Scaife, R.; El-Hofy, H. Abrasive Water Jet Machining of Multidirectional CFRP Laminates. *Procedia CIRP* **2018**, *68*, 535–540. [CrossRef]

17. Schwartzentruber, J.; Spelt, J.K.; Papini, M. Prediction of surface roughness in abrasive waterjet trimming of fiber reinforced polymer composites. *Int. J. Mach. Tools Manuf.* **2017**, *122*, 1–17. [CrossRef]

18. Yuvaraj, N.; Kumar, M.P. Cutting of aluminium alloy with abrasive water jet and cryogenic assisted abrasive water jet: A comparative study of the surface integrity approach. *Wear* **2016**, *362–363*, 18–32. [CrossRef]

19. MM, I.W.; Azmi, A.; Lee, C.; Mansor, A. Kerf taper and delamination damage minimization of FRP hybrid composites under abrasive water-jet machining. *Int. J. Adv. Manuf. Technol.* **2018**, *94*, 1727–1744. [CrossRef]

20. Alberdi, A.; Suárez, A.; Artaza, T.; Escobar-Palafox, G.A.; Ridgway, K. Composite Cutting with Abrasive Water Jet. *Procedia Eng.* **2013**, *63*, 421–429. [CrossRef]

21. Unde, P.D.; Gayakwad, M.D.; Patil, N.G.; Pawade, R.S.; Thakur, D.G.; Brahmankar, P.K. Experimental Investigations into Abrasive Waterjet Machining of Carbon Fiber Reinforced Plastic. *J. Compos.* **2015**, *2015*, 1–9. [CrossRef]

22. Ravi Kumar, K.; Sreebalaji, V.S.; Pridhar, T. Characterization and optimization of Abrasive Water Jet Machining parameters of aluminium/tungsten carbide composites. *Meas. J. Int. Meas. Confed.* **2018**, *117*, 57–66. [CrossRef]

23. Phapale, K.; Singh, R.; Patil, S.; Singh, R.K.P. Delamination Characterization and Comparative Assessment of Delamination Control Techniques in Abrasive Water Jet Drilling of CFRP. *Procedia Manuf.* **2016**, *5*, 521–535. [CrossRef]

24. Schwartzentruber, J.; Spelt, J.K.; Papini, M. Modelling of delamination due to hydraulic shock when piercing anisotropic carbon-fiber laminates using an abrasive waterjet. *Int. J. Mach. Tools Manuf.* **2018**, *132*, 81–95. [CrossRef]

25. Schwartzentruber, J.; Papini, M. Abrasive waterjet micro-piercing of borosilicate glass. *J. Mater. Process. Technol.* **2015**, *219*, 143–154. [CrossRef]

26. Ahmed, T.M.; El Mesalamy, A.S.; Youssef, A.; El Midany, T.T. Improving surface roughness of abrasive waterjet cutting process by using statistical modeling. *CIRP J. Manuf. Sci. Technol.* **2018**, *22*, 30–36. [CrossRef]

27. Ming Ming, I.W.; Azmi, A.I.; Chuan, L.C.; Mansor, A.F. Experimental study and empirical analyses of abrasive waterjet machining for hybrid carbon/glass fiber-reinforced composites for improved surface quality. *Int. J. Adv. Manuf. Technol.* **2018**, *95*, 3809–3822. [CrossRef]

28. Gnanavelbabu, A.; Saravanan, P.; Rajkumar, K.; Karthikeyan, S. Experimental Investigations on Multiple Responses in Abrasive Waterjet Machining of Ti-6Al-4V Alloy. *Mater. Today Proc.* **2018**, *5*, 13413–13421. [CrossRef]

29. Prasad, K.S.; Chaitanya, G. Selection of optimal process parameters by Taguchi method for Drilling GFRP composites using Abrasive Water jet machining Technique. *Mater. Today Proc.* **2018**, *5*, 19714–19722. [CrossRef]

30. Shanmugam, D.K.; Nguyen, T.; Wang, J. A study of delamination on graphite/epoxy composites in abrasive waterjet machining. *Compos. Part A Appl. Sci. Manuf.* **2008**, *39*, 923–929. [CrossRef]

31. Mayuet, P.F.; Girot, F.; Lamíkiz, A.; Fernández-Vidal, S.R.; Salguero, J.; Marcos, M. SOM/SEM based Characterization of Internal Delaminations of CFRP Samples Machined by AWJM. *Procedia Eng.* **2015**, *132*, 693–700. [CrossRef]

32. Shukla, R.; Singh, D. Experimentation investigation of abrasive water jet machining parameters using Taguchi and Evolutionary optimization techniques. *Swarm Evol. Comput.* **2017**, *32*, 167–183. [CrossRef]

33. Gupta, V.; Pandey, P.M.; Garg, M.P.; Khanna, R.; Batra, N.K. Minimization of Kerf Taper Angle and Kerf Width Using Taguchi's Method in Abrasive Water Jet Machining of Marble. *Procedia Mater. Sci.* **2014**, *6*, 140–149. [CrossRef]

34. Dhanawade, A.; Kumar, S. Experimental study of delamination and kerf geometry of carbon epoxy composite machined by abrasive water jet. *J. Compos. Mater.* **2017**, *51*, 3373–3390. [CrossRef]

![materials logo] *materials*

MDPI

Article

From Design to Manufacture of a Carbon Fiber Monocoque for a Three-Wheeler Vehicle Prototype

Alessandro Messana, Lorenzo Sisca, Alessandro Ferraris, Andrea Giancarlo Airale, Henrique de Carvalho Pinheiro, Pietro Sanfilippo and Massimiliana Carello *

Department of Mechanical and Aerospace Engineering, Politecnico di Torino, 10129 Turin, Italy; alessandro.messana@polito.it (A.M.); lorenzo.sisca@polito.it (L.S.); alessandro.ferraris@polito.it (A.F.); andrea.airale@polito.it (A.G.A.); henrique.decarvalho@polito.it (H.d.C.P.); pietro.sanfilippo1990@gmail.com (P.S.)
* Correspondence: massimiliana.carello@polito.it; Tel.: +39-011-0906946

Received: 18 December 2018; Accepted: 16 January 2019; Published: 22 January 2019

Abstract: This paper describes the design process of the monocoque for IDRAkronos, a three-wheeler hydrogen prototype focused on fuel efficiency, made to compete at the Shell Eco-Marathon event. The vehicle takes advantage of the lightweight and high mechanical performance of carbon fiber to achieve minimal mass and optimized fuel consumption. Based on previous experiences and background knowledge, the authors describe their work toward a design that integrates aerodynamic performance, style, structural resistance and stiffness. A portrayal of the objectives, load cases, simulations and production process—that lead to a final vehicle winner of the Design Award and 1st place general at the 2016 competition—is presented and discussed.

Keywords: carbon fiber; structural analysis; monocoque structure; lightweight design; low consumption vehicle; three-wheeler vehicle

---

1. Introduction

The importance of lightweight design and high mechanical performance in contemporary automotive development is undisputed. Certainly, one of the main related goals is fuel efficiency.

In order to achieve a good compromise between mass, mechanical properties, aerodynamic performance and style, the use of carbon fiber stands out. Its application allows creative shapes to couple with elevated engineering performance, permitting high design freedom.

The main issues of carbon fiber technology are the cost of production and the recyclability; therefore, its major applications still lie in the racing field and niche markets.

This paper presents the design and manufacturing of a three-wheeler prototype for the Shell Eco-marathon competition [1], which awards the vehicle that obtains the lowest fuel consumption measured in terms of km, with a liter equivalent fuel (calculated by Shell, using chemical formulations).

Starting from experience in other vehicles designs [2–4] at the Politecnico di Torino and also by other teams [5], the new vehicle—IDRAkronos—has been developed, using a carbon fiber monocoque to reach minimal mass, while maintaining structural resistance. The prototype has three wheels; two front steering wheels (covered by the body to reduce aerodynamic drag) and one rear wheel powered by a brushed electric motor, supplied by a hydrogen fuel cell. The upper limit of 40 kg has been respected, including by all the sub-systems; steering, breaks, wheels, cockpit, electric wiring, controls, fuel cell, the electric motor, and transmission.

During the thorough technical inspection that takes place before the competition [1], one of the most challenging tests for the vehicle body was the application of a 70 kg load on the highest point of the vehicle (roll bar). The structure of the vehicle should not deform permanently, nor show ruptures during the test. To withstand this load, a good technological solution is the carbon fiber composite, which has excellent performance in terms of low density and mechanical properties [6,7], allowing us to attain minimum mass and high resistance, but also to make an aerodynamically optimized shape.

Apart from the technical regulation constrains and some geometrical fixed features (for instance, the vehicle track) the design followed was conducted in order of CAD (Computer Aided Design), FEM (Finite Element Method) analysis, CFD (Computational Fluid Dynamics) analysis and some iteration between structural resistance and aerodynamic performance. Once the final shape was achieved, a definitive FEM model was developed, concurrently with the ply book for the manufacturing process.

A typical manufacturing technology used to produce carbon fiber components for automotive applications is the use of pre-pregs, which are fabric composite materials already littered with resin. The pre-pregs are shaped manually in molds and then polymerized in an autoclave, using a vacuum bag around the mold. Unidirectional and multi directional carbon fiber fabrics were used, with a specific ply lay-up to balance the stress due to pressure and temperature variation, which occurs during the curing cycle in the autoclave. By doing so, it is possible to obtain components with elevated uniformity and performance.

2. Materials and Methods

2.1. Geometry Design

A significant change in IDRAkronos' layout (Figure 1)—with respect to the prototypes designed previously [2–4]—is the location of the rear wheel behind the firewall, and the fuel cell with the hydrogen tank at the end of the car. This solution grants not only a considerably stiffer rear assembly and a shorter wheelbase, but also higher design freedom for the external shape of the vehicle's tail. Moreover, this position enhances the ventilation of the fuel cell, increasing the powertrain's efficiency.

Figure 1. IDRAkronos' layout.

The position of the driver was set such that the steering bar passed below the knees, while the wheels were leveled with the hip. The choice of the vehicle's height was the result of a compromise, where on the one hand, a low height would reduce the frontal area (and therefore the aerodynamic drag) and avoid rollover, but a more elevated floor would prevent the downforce caused by ground effects, thus reducing the rolling resistance.

The engine was directly mounted on the chassis in order to minimize relative motion between the pinion and the gear mounted on the wheel. This way, a better coupling efficiency was achieved during cornering, where the components typically tend to move apart.

Moreover, IDRAkronos had three particularly well-developed characteristics:

1. Covered front wheels that reduced turbulence generated by the spinning motion. In this case, the wheel arch volume had to include the envelope of the steering wheels, increasing the frontal area of the car, which was compensated for, by a significant reduction in the drag coefficient.

2. Reduced wheelbase of the vehicle helps decreased the frontal area of the car and also the volume of the wheel arches, improving the aerodynamic resistance (a lower wheelbase means a lower steering angle is needed to run the same curve). Another advantage of the reduced wheelbase is that the length of the structural part of the car was shortened, allowing a reduction of the bending stresses on the monocoque, and thus helping to achieve a lighter vehicle.

3. Increased length of the body allowed us to close the tail with a smaller angle, delaying the transition from laminar to turbulent flow as far back as possible, reducing the dimensions of the wake, and therefore the drag coefficient.

The software used to design the surfaces is Alias (2015 Version) by Autodesk, San Rafael, CA, USA. The design of IDRAkronos (Figure 2) was influenced by profile 4415 of NACA (National Advisory Committee for Aeronautics) airfoil, with dimensions according to their characteristics and the compatibility with the proportions of the prototype. The NACA duct was designed to work with a rear opening which allows the air inside the rear compartment to flow outside, avoiding the spoon effect. The refrigeration of the fuel cell was one of the main objectives; in particular, the temperature profile was estimated on its duty cycle during the race, at different times of the day. The temperature gap between the two solutions justified the presence of the NACA duct in order to improve the powertrain's efficiency.

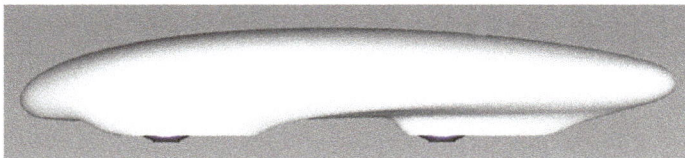

Figure 2. IDRAkronos' external shape.

The body of IDRAkronos is subdivided into the following parts (Figure 3):

- The front end
- The central part
- The rear end
- The top cover

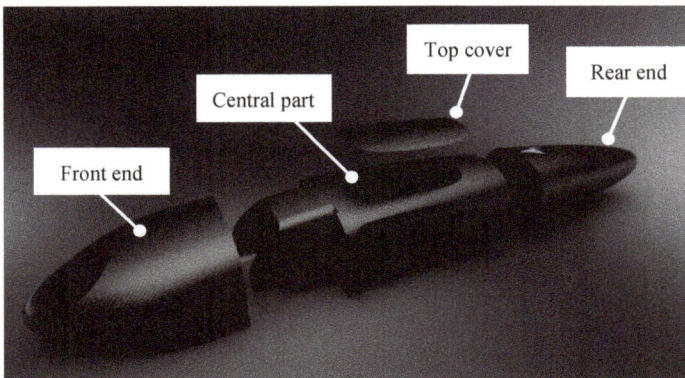

Figure 3. Subdivisions of the IDRAkronos' body.

The front end, rear end, and the cover did not add a relevant contribution to the stiffness of the vehicle, and the central part had the purpose of withstanding all static and dynamic loads that acted on the vehicle.

With IDRAkronos, the objective was to have a closed cross section in order to significantly improve the moment of inertia of the monocoque, improving the stiffness due to the geometry factor, and therefore allowing us to use less material to achieve the same displacements.

2.2. Materials

The monocoque of IDRAkronos is made of 0.25 mm thick sheets of 2×2 twill T300 carbon fiber/epoxy pre-preg, layered in a sandwich structure within the Nomex honeycomb. The mechanical properties of these materials are reported in Table 1. Carbon fiber based composites show a high strength-to-volume ratio, which make them ideal for applications that need both lightness and high strength. Preferred orientations used for pre-preg sheets were $0°$, $45°$, $-45°$, and $90°$, while avoiding $30°$ and $60°$ for simplicity during the manufacturing process.

Table 1. Mechanical properties of carbon fiber laminate and Nomex honeycomb.

Property	T300/Epoxy Composite	Nomex Honeycomb
E—Young's Modulus (GPa)	57	0.9
G—Shear Modulus (GPa)	3	0.3
σ_t—Tensile Strength (MPa)	570	-
σ_c—Compressive Strength (MPa)	530	80
ν—Poisson's Ratio	0.05	0.4
ϱ—Specific Mass (kg/dm^3)	1.4	0.06

The sandwich structure is frequently used in the motorsport and aerospace industry due to its ability to combine the best properties of both materials. Nomex honeycomb is a core material made up of aramid polyamide fibers, which have low specific mass and high torsional resistance, impact toughness and vibration damping. The possibility of using a core material with very low density allowed us to employ a higher thickness core (ranging from 6 mm to 13 mm in this application, as shown in Figure 10) and, importantly, to increase the moment of inertia of the section without adding too much mass.

2.3. FEM Model

Linear static analyses of IDRAkronos were made using the Hypermesh and Hyperview software (v.14) and using the Optistruct solver (both provided by Altair, Troy, MI, USA). Figure 4 shows the CAD design of the body and its smaller subdivisions, with different sizes and curvature used to better analyze and control the FEM mesh quality. In the pre-processing interface, the geometry clean-up and meshing process was carefully carried out, as depicted in Figure 5.

Figure 4. Central part CAD design (**left**) and subdivision to enhance mesh quality (**right**).

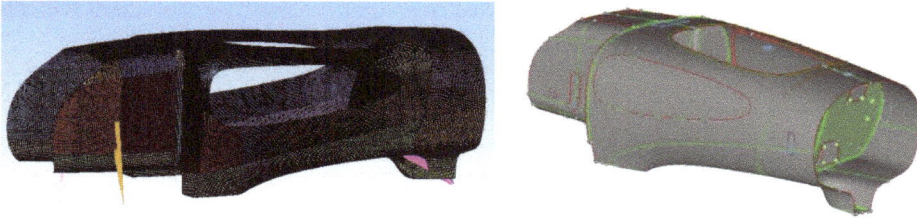

Figure 5. FEM shell elements mesh definition (**left**) and pre-processing geometry cleanup (**right**).

Two-dimensional elements have been used for the mesh, since one dimension is negligible with respect to the others for the panels, under the hypothesis that the distribution of stresses along the axis orthogonal to the element plane is irrelevant. Although stiffer than quadrilateral elements, triangular elements were necessary to follow the fairly complex geometry. The quality of an element was determined by minimum length, skew angle, Jacobian, aspect ratio, and warpage. In fact, reducing the size of the elements increased their number, so results were more accurate with a longer computational time.

MAT 8 was used in Hypermesh to define linear temperature-independent orthotropic materials for two-dimensional elements. The PCOMPP card was used in combination with the STACK and PLY cards to create composite properties through the Ply-based composite definition.

Composite materials are highly anisotropic and behave differently based on their orientation. The 0° direction was oriented along the x-axis of the vehicle.

The track on which the race was held was analyzed and, due to the speed profile expected during the race, a target on the limit for rollover conditions was set, assuming infinite adherence from the tires. From the Shell Eco-marathon regulations [1], the vehicle was required to be able to run a 10 m radius curve at a speed of up to 27 km/h. Multibody dynamic simulations made with Adams View confirmed the above requirements, outputting the forces acting on the wheels right before rollover (Table 2).

Table 2. Rollover forces.

Tire	F_z [N]	F_y [N]
External front	650	350
Internal front	0	0
Rear	350	150

In regards to the other loads, the mass of the head was applied in a rectangular shape onto the firewall, since the head was placed above the fire extinguisher-holding structure. The loads on the back and hips were applied on an area with a width roughly equal to the shoulder width of the driver, and the loads caused by the driver's feet were applied on the front end of the vehicle. Constraints, similar to those on the load step were placed, which assured that the driver was inside the vehicle.

Furthermore, the firewall had the function of a roll bar, which had to be capable of withstanding a static load of 700 N, as the regulations stated.

3. Results

3.1. FEM Results

The design phase was composed of three principal steps; free-size-optimization, discrete-size-optimization, and shuffle-optimization. As the monocoque of IDRApegasus (the predecessor of IDRAkronos) never reached more than 18 mm in thickness, the maximum thickness of the laminate was set to 20 mm during optimization. Total displacement response to the load steps "driver inside, stand, and exit" was considered to be 1 mm. For most parts, two plies of carbon fiber for each orientation were more than enough, and the minimum thickness for each orientation was, in fact, set to 0.5 mm, corresponding to two plies, in order to keep the laminate balanced and guarantee a certain degree of isotropy.

To improve correlation with the real application and reduce to bare minimum the amount of material used to obtain the preset target in terms of stiffness, the contribution of the side windows was computed inside the FEM model. This is an essential step of the design and optimization process, since an inaccurate description of the general behavior of the structure could lead to the exaggeration of the thickness (and therefore weight) of the body.

For optical reasons, 3 mm thick scratch-resistant antireflection-certified panels made by Lexan were used, resulting in a 1.5 kg mass addition to the vehicle. The lateral windows' contribution to the bending stiffness of the monocoque was very important, as depicted by Figure 6, which shows the effect on the roll-over conditions, and Figure 7, which shows the effect on the displacement due to the driver's hold during the exit from the vehicle. It is clear, after these results, that the inclusion of the glass contributed majorly to the structural performance, allowing a proper setup for the optimization and demonstrating that these elements should be carefully designed and installed.

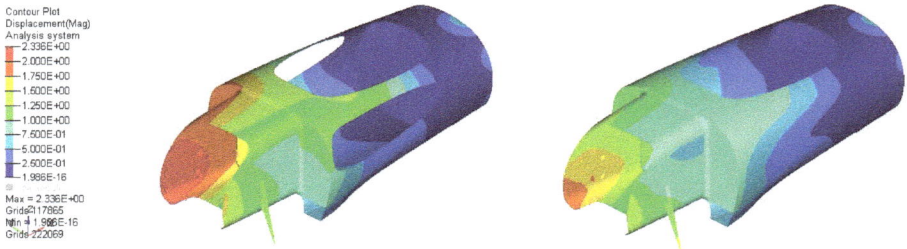

Figure 6. Rollover displacements without (**left**) and with (**right**) windows.

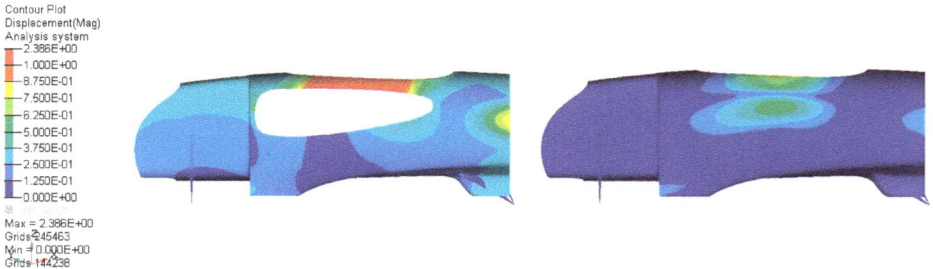

Figure 7. Exit displacements without (**left**) and with (**right**) windows.

The same approach has been used for the design of the front end and the tail of the vehicle, with lower loads.

Given strict requirements from the displacement point of view, stresses remained very low, of the order of few MPa, as shown in Figure 8. Nonetheless, for the sake of completeness, a ply failure check was computed using Tsai Wu theory for composite materials. As expected, ply failure never occurred in any load step, even in the most severe tests of the monocoque.

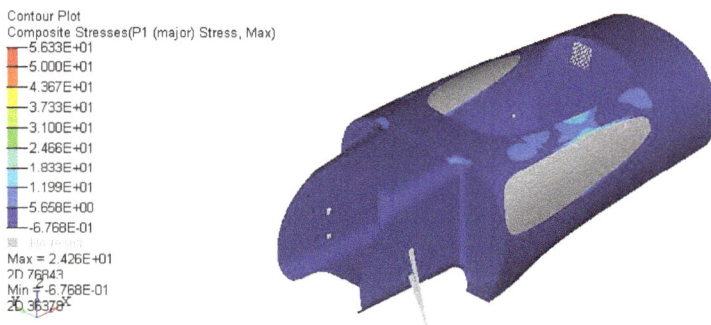

Figure 8. Stress distribution during the exit maneuver.

Many analyses, such as the free-size-optimization (Figure 9), were carried out in order to define the best shapes and thicknesses of the honeycomb inside the body. This was extremely important in order to achieve the goal of minimum mass (and ultimately lower consumption) for the vehicle, while maintaining mechanical performance. Bearing in mind that the limits on the thicknesses available were only 13 mm, 6 mm, and 3 mm, and the shapes that could be obtained due to different curvatures of the monocoque, the final lamination ply book is reported in Figure 10.

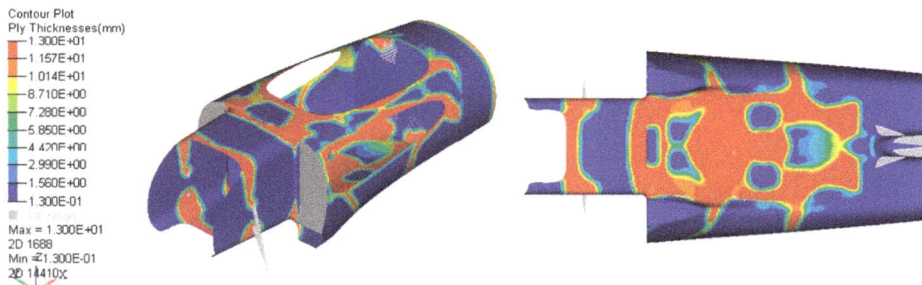

Figure 9. Free-size-optimization of honeycomb thickness: Top (**left**) and bottom (**right**) view.

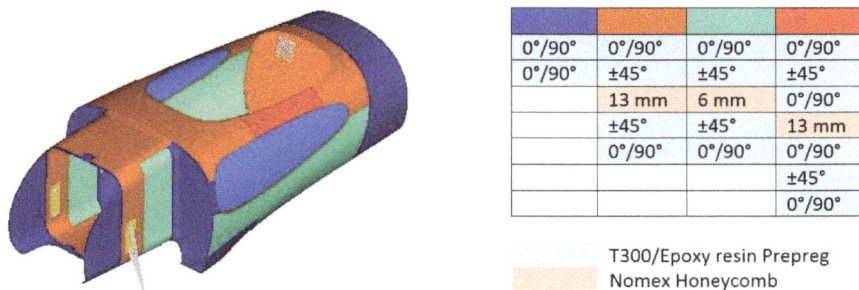

0°/90°	0°/90°	0°/90°	0°/90°
0°/90°	±45°	±45°	±45°
	13 mm	6 mm	0°/90°
	±45°	±45°	13 mm
	0°/90°	0°/90°	0°/90°
			±45°
			0°/90°

T300/Epoxy resin Prepreg
Nomex Honeycomb

Figure 10. Final laminate (**left**) and stacking sequence (**right**).

3.2. Production Process

The final properties of the components in composite materials, other than the properties of the carbon fiber and matrix, depend strongly on the production process. For this reason, it was not enough to choose the right type of materials, it was also of great importance to evaluate the appropriate production process in order to guarantee good quality of the final product. Inter-laminar cohesion, for example, was strictly dependent on the absence of air bubbles absorbed during the manufacturing of the laminate.

The most suitable process to realize the components of the prototype was the use of the autoclave vacuum bag, which required a phase of composite lamination in a mold and a phase of thermal consolidation both in vacuum and under pressure.

Blocks of epoxy resin (Figure 11) RAKU-TOOL WB0691 (Rampf Company, Gräfenberg, BY, Germany) were used to reduce the costs of metal. Resin male molds were used to realize carbon fiber female molds, which carried out the first carbon fiber lamination. Figure 12, on the left, shows the closure of the female molds in the release film, and Figure 12, on the right, shows the closure of the breather film and vacuum bag before the autoclave cycle. The autoclave cycle used lasted 3 h, at 130 °C and 5.5 bar.

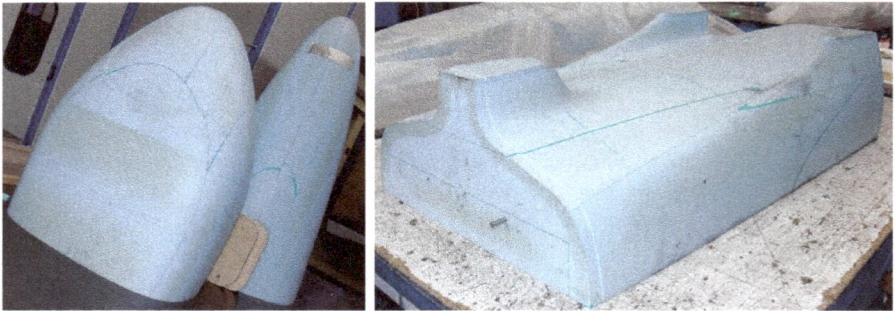

Figure 11. Male resin molds; front and rear end (**left**), and central part (**right**).

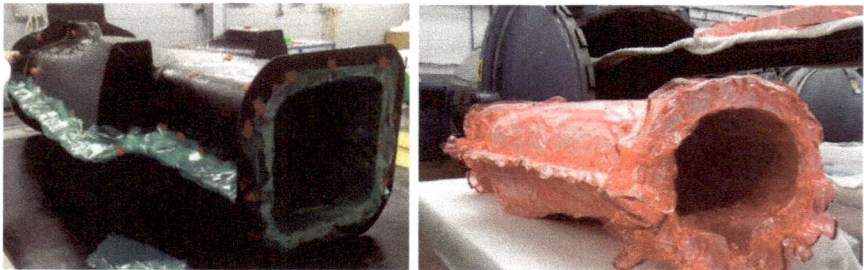

Figure 12. First lamination: Release film (**left**) and vacuum bag (**right**) on the female central mold.

For the second lamination, honeycomb sheets were glued in the central part of the body, only where designed by FEM (Figure 13, left). The composite layers were laminated to realize the internal shell (Figure 13, right). In order to avoid the collapse of Nomex, the second curing cycle was done at a maximum of 1.5 bar.

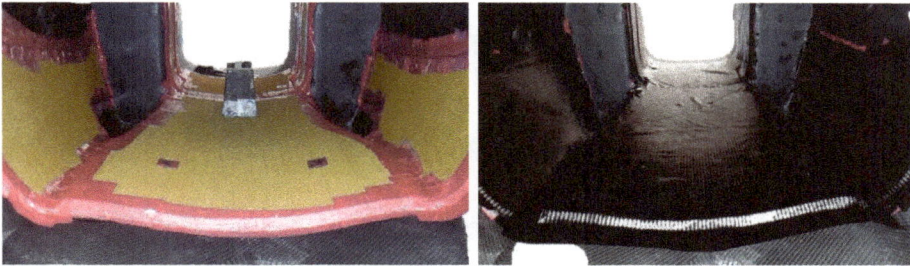

Figure 13. Second lamination in the central part: Nomex honeycomb (**left**) and composite layers (**right**).

After the body was released from the mold (Figure 14), some trimming was done on the tunnel, the windows, and the wheel mounting zone. To obtain a perfectly vertical laminate needed to mount the wheel, six of 22 plies of carbon fibers were sacrificed in the milling machine. Before starting the trim, the body was properly positioned and the symmetry plane was traced. This was crucial, so as to have the faces, on which the wheel brackets had to be mounted, perfectly parallel to the vehicle's z–x plane, with an offset of ±210 mm along the y- axis, and therefore to minimize the error on the final wheel alignment. Finally, all the holes were drilled, the metal inserts screwed in, and the hooks affixed on the monocoque.

Figure 14. Final central part.

The manufacture of the firewall was relatively simple, with only one curing cycle of 12 mm of honeycomb sheets between carbon fiber plies. To properly glue the firewall on the monocoque, a guide was built with four centering pins placed on the main body to guarantee the perfect placement of the firewall at exactly 15° from the vertical plane. An error at this stage could have jeopardized the position of the rear wheel, both in relative misalignment and in absolute positioning, leading to a permanent, unwanted pitch or roll angle of the whole vehicle.

The front end was laminated with two plies, while four plies were placed in the frame in contact with the main body, to increase the stiffness and the handling. The rear end was laminated with two plies, and after curing, the only operation made on the rear end was the cutting of a hole used to glue the NACA duct, which was 3D printed previously.

4. Discussion

Mass reduction is a key factor in the design of a prototype vehicle aimed at a low consumption competition, such as the Shell Eco-marathon. In this paper, the structural design of IDRAkronos prototype's body, from the concept phase (Figure 3) to the manufacturing process (Figure 15) was illustrated. Trying to minimize the mass of the vehicle, the authors highlighted the main steps of the CAD design that creates the general shape, the aerodynamic requirements, and finally the FEM simulations and optimization process. In this way, the ply book of the composite was defined before fabricating the body using the pre-preg, hand lay-up, and autoclave production processes.

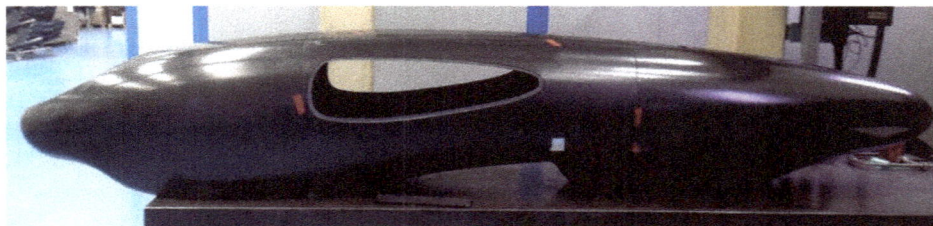

Figure 15. Full vehicle body assembled.

The design of the carbon fiber monocoque delivered great results in terms of mass and stiffness. The technical solutions adopted were proved to be successful, and each detail was important in achieving the desired mechanical performance of the structure, and therefore the general objectives for the whole vehicle. The main body had a mass of 7.5 kg, a tail mass of 1.2 kg, while the front-end mass was 2.5 kg. This lightweight design contributed to obtaining a total vehicle mass of just 39 kg.

The design methodology and insights provided by this paper may contribute to the development of similar low consumption vehicles, creating a guideline for the application of carbon fiber composite materials in monocoque structures [8–10], or any other component with similar complexity and variety of constrains and goals.

Author Contributions: Conceptualization, P.S. and M.C.; methodology, A.M. and L.S.; software, A.M., L.S. and P.S.; validation, A.F., A.G.A.; formal analysis, A.F., A.G.A., H.d.C.P.; investigation, L.S.; data curation, L.S. and A.M.; writing—original draft preparation, M.C. and L.S.; writing—review and editing, M.C. and H.d.C.P.; supervision, M.C.; project administration, M.C.; funding acquisition, M.C.

Funding: This research was funded by Politecnico di Torino—"Commissione contributi e progettualità studentesca" and all the sponsor and partner of Team H$_2$politO (www.polito.it/h2polito).

Acknowledgments: The authors would like to thanks: Altair Italia s.r.l. for the software licenses and all the students of Team H$_2$politO.

Conflicts of Interest: The authors declare no conflict of interest.

References

1. Shell Eco-Marathon, Official Rules. Available online: https://www.shell.com/energy-and-innovation/shell-ecomarathon/europe/for-europe-participants.html (accessed on 18 January 2019).
2. Airale, A.G.; Carello, M.; Scattina, A. Carbon fiber monocoque for a hydrogen prototype for low consumption challenge. *Materalwiss. Werkstofftech.* **2011**, *42*, 7. [CrossRef]
3. Carello, M.; Airale, A.G.; Messana, A. IDRApegasus: A carbon fiber monocoque vehicle prototype. *Materalwiss. Werkstofftech.* **2014**, *45*, 9. [CrossRef]
4. Carello, M.; Messana, A. IDRApegasus: A fuel-cell prototype for 3000 km/L. *Comput.-Aided Des. Appl.* **2015**, *12*, 56–66. [CrossRef]
5. Tsirogiannis, E.C.; Siasos, G.I.; Stavroulakis, G.E.; Makridis, S.S. Lightweight Design and Welding Manufacturing of a Hydrogen Fuel Cell Powered Car's Chassis. *Challenges* **2018**, *9*, 25. [CrossRef]
6. Fasana, A.; Ferraris, A.; Berti Polato, D.; Airale, A.G.; Carello, M. Composite and Damping Materials Characterization with an Application to a Car Door. In *Mechanisms and Machine Science, Proceedings of the International Conference of IFToMM ITALY: Advances in Italian Mechanism Science, Cassino, Italy, 29–30 November 2018*; Springer: Cham, Switzerland, 2019; Volume 68, pp. 174–184. [CrossRef]
7. Carello, M.; Airale, A.G.; Ferraris, A.; Messana, A.; Sisca, L. Static Design and Finite Element Analysis of Innovative CFRP Transverse Leaf Spring. *Appl. Compos. Mater.* **2017**, *24*, 16. [CrossRef]

8. Weidner, L.R.; Radford, D.W.; Fitzhorn, P.A. A Multi-Shell Assembly Approach Applied to Monocoque Chassis Design. In Proceedings of the Motorsports Engineering Conference & Exhibition, Indianapolis, IN, USA, 2 December 2002. [CrossRef]

9. Wu, J.; Agyeman Badu, O.; Tai, Y.; George, A.R. Design, Analysis, and Simulation of an Automotive Carbon Fiber Monocoque Chassis. *SAE Int. J. Passeng. Cars Mecha. Syst.* **2014**, *7*, 838–861. [CrossRef]

10. Tsirogiannis, E.C.; Stavroulakis, G.E.; Makridis, S.S. Design and Modelling Methodologies of an Efficient and Lightweight Carbon-fiber Reinforced Epoxy Monocoque Chassis, Suitable for an Electric Car. *Mater. Sci. Eng. Adv. Res.* **2017**, *2*, 5–12. [CrossRef]

MDPI

St. Alban-Anlage 66

4052 Basel

Switzerland

Tel. +41 61 683 77 34

Fax +41 61 302 89 18

www.mdpi.com

Materials Editorial Office

E-mail: materials@mdpi.com

www.mdpi.com/journal/materials

www.ingramcontent.com/pod-product-compliance
Lightning Source LLC
Chambersburg PA
CBHW051855210326
41597CB00033B/5908